Hydraulic Design and Management of Wastewater Transport Systems

Hydraulic Design and Management of Wastewater Transport Systems

CAPWAT Manual

Michiel Tukker
Kees Kooij
Ivo Pothof

IWA PUBLISHING

Published by **IWA Publishing**
Alliance House
12 Caxton Street
London SW1H 0QS, UK
Telephone: +44 (0)20 7654 5500
Fax: +44 (0)20 654 5555
Email: publications@iwap.co.uk
Web: www.iwapublishing.com

This edition published 2016

ISBN: 9781780407814 (eBook)
ISBN: 9781780407821 (Paperback)

This is a special edition of the book "Hydraulic design and management of wastewater transport systems" that was first published by Deltares in 2013.

Deltares is an independent Dutch research institute for water, subsurface and infrastructure issues. Its mission is to develop and apply top-level expertise to these issues for people, planet and prosperity. Worldwide, around 900 staff are working to find innovative solutions that make living in the world's deltas, coastal areas and river basins safe, clean and sustainable.

Table of Contents

List of symbols

Symbol	Unit	Meaning
a	-	Regression constant
A	m^2	Cross-section
A_v	m^2	Valve-specific discharge coefficient
$A_{w,c}$	m^2	Wet cross-section at critical water depth
$A_{w,n}$	m^2	Wet cross-section at normal water depth
b	m	Width
$b_{s,c}$	m	Width at water surface at critical water depth
c	m/s	Speed of pressure wave
c_{gas}	m/s	Pressure wave speed with gas bubbles in the system
c_s	m^3/m^3	Saturation concentration
C	J	Energy of surge vessel
C_D	-	Drag coefficient
C_v	$USgal(psi)^{0.5}$	Valve-specific discharge coefficient
D	m	Bubble diameter
D	m	Diameter
D_N	m	Hydraulic diameter of water film at normal depth
DN	mm	Nominal diameter
e	m	Wall thickness
E	N/m^2	Elasticity modulus
E	m	Energy head

E_{pot}	J	Potential energy
E_{kin}	J	Kinetic energy
f	Hz	frequency
F	-	Flow number
F_w	-	Flow number (for water)
F_g	-	Flow number (for the gas flow rate)
F_p	-	Piping geometry factor
g	m/s^2	Gravitational acceleration
h	m	Pressure head
h_e	m	Final water depth
$\Delta h_{c,l}$	m	Pressure difference at which the check valve begins to open
H	m	Energy head
H	m	Outflow height
H_p	m	Static pump head
ΔH_{dyn}	m	Dynamic energy loss
ΔH_{gas}	m	Energy loss due to gas volume
$\Delta H \bullet$	-	Dimensionless energy loss
k	-	Polytropic or Laplace coefficient
k	m	Wall roughness
K_v	$m^3/h(bar)^{0.5}$	Valve-specific discharge coefficient
K	N/m^2	Compressibility modulus
L	m	Length of deflection plate
L	m	Length of pipeline
L_B	m	Bubble length
L_k	m	Characteristic length
L_{jx}	m	Length of bubble plume
$L_{max,hor}$	m	Maximum bubble length
M	kg	Mass
N	rpm	Pump speed
Δp	Pa	Pressure difference
O	m	Hydraulic circumference

P	Pa	Pressure
P	W	Power
P_v	Pa or bara	Vapour pressure
Q_{dwf}	m^3/s	Dry-weather design flow rate
$Q_{g,in}$	m^3/s	Influent air flow rate
$Q_{g,t}$	m^3/s	Transported air flow rate
Q_G	m^3/s	Gas flow rate
Q_L	m^3/s	Liquid flow rate
Q_w	m^3/s	Water flow rate
R	m	Radius
Re	-	Reynolds number
S	m	Distance between pipe exit and deflection plate
T	s	Time
Δt_R	s	Duration of return flow velocity through check valve
T	°C	Temperature
V	m/s	Flow velocity
v_0	m/s	Critical velocity at which check valve begins to open
v_e	m/s	Exit velocity
v_{min}	m/s	Minimum velocity for check valve in operating range
v_R	m/s	Return flow velocity through check valve
V	m^3	Volume
V_D	m/s	Flow velocity at bell mouth
X	m	Horizontal distance
Y	m	Vertical distance
y_c	m	Water layer thickness near bend
y_{min}	m	Critical depth
Z	m	Elevation
α	°	Slope angle
λ	-	Darcy-Weisbach friction factor
μ	-	Flow contraction coefficient
μ	s	Reflection time of pressure wave
ν	m^2/s	Kinematic viscosity

η	Ns/m^2	Dynamic viscosity
ρ_G	kg/m^3	Gas density
ρ_L	kg/m^3	Liquid density
σ	N/m	Surface tension
σ	N/m^2	Tangential wall tension
θ	%	Valve position
τ	N/m^2	Shear stress
$\xi_{i,loc}$	-	Energy loss coefficients for local losses

About this Manual

This manual was compiled by a group of hydraulic specialists at Deltares: ing. M. Tukker, ing. C. Kooij and Dr. Ir. I. Pothof (PhD). The authors were supported by reviewers and a technical committee composed by the following members:

- F. Clemens (Deltares/TU Delft)
- R. v.d. Anker (Waterschap Rivierenland)
- J. Driessen (Grontmij)
- M. Geise (Xylem, previously ITT Water & Wastewater)
- J. Kranendonk (Gemeentewerken Rotterdam)
- C. Lubbers (Royal HaskoningDHV)
- P. v. Rosmalen (Hoogheemraadschap Delfland)
- F. v. Zijl (Waterschap Brabantse Delta)

This manual is the translation of the final submittal of the research project CAPWAT-II that was carried out by Deltares and TU Delft/CiTG in the period between 2007 and 2010. After a first Dutch version released in June 2010, comments have been received and incorporated into subsequent releases. In 2010 and 2011, the Dutch applied research foundation Stowa commissioned the organisation of several workshops to transfer knowledge and to stimulate using this manual. Besides that, knowledge distribution has taken place via specific mini-courses of RIONED. Subsequently, the second Dutch edition was published in March 2012. The following organizations participated in the CAPWAT research project:

- Hoogheemraadschap Delfland
- Hoogheemraadschap Hollands
- Noorderkwartier
- Waterschap Brabantse Delta
- Waterschap Hollandse Delta
- Waterschap Reest en Wieden
- Waterschap Rivierenland
- Waterschap Zuiderzeeland
- Wetterskip Fryslân
- Waternet
- Gemeentewerken Rotterdam
- Grontmij
- Royal HaskoningDHV
- ITT Water&Wastewater
- Aquafin
- STOWA

The authors are grateful to the technical committee, reviewers and supporting organizations for their commitment and valuable contributions. Without their input, this manual would not have reached the high status and recognition in the Dutch wastewater business.

On behalf of Deltares,

Bas van Vossen
Head of department Industrial Hydrodynamics-Deltares

Chapter 1
Foreword

1.1 NECESSITY OF THE MANUAL

The creation of the manual 'Hydraulic design and management of wastewater transport systems' arose from the research project CAPWAT (CAPacity loss in wasteWATer pressure pipelines), which researched the mechanisms for the creation, stagnation and discharge of gas bubbles in wastewater pressure pipelines. During this six-year research programme, it was recognised that there is no hydraulic manual/guideline that focuses on the entire wastewater pressure pipeline system, the processes it includes, and the interaction between the pressure pipeline and the pumping station.

Processes that hardly or never occur in clean water transport systems (such as cooling water, drinking water) must be taken into account when designing a wastewater transport system. In case of wastewater, we have to deal with discontinuous supply. The type of sewer system (combined, separated, improved separated) determines the distribution of supply flow during both dry and wet weather.

The characteristics of the wastewater (such as surface tension and turbidity) vary in time and per location, as well as the waste load (floating and non-floating parts).

Usually, the designers and managers are used to thinking in terms of stationary processes. A wastewater pressure pipeline does not operate according to a stationary process, certainly not in dry weather conditions. Knowledge about the dynamic processes (variation in time) that occur in a wastewater transport system is necessary in order to determine the design and management guidelines.

Two important processes that the designer/manager deals with are:

- The process of the creation, stagnation and transportation of gas bubbles, and
- The water hammer phenomenon.

Another aspect is that the wastewater transport system is becoming more complex. Due to building larger sewage water treatment plants, wastewater is being transported over greater distances and increasingly more (and smaller) pipelines connect to the main sewers. The operation of the pumping stations is largely determined by how the entire system behaves. Insight into this operation is, therefore, crucial for proper design and management.

The manual 'Hydraulic design and management of wastewater transport systems' provides an overview of all the aspects and interrelatedness that are crucial for the hydraulic design and management of a wastewater transport system. A wastewater transportation system is understood to mean the pressure pipeline and/or the pressure pipeline system, including the pumping station and the receiving basin.

The central point of the design is to create an independent and safe system with the necessary transport capacity at minimum societal costs. Predominantly, the management aspect focuses on guidelines to maintain the design principles regarding capacity and required energy.

1.2 SCOPE OF THE MANUAL

The purpose of this manual is to create a compilation of all the hydraulic knowledge that is necessary for designing a wastewater transport system and to manage it operationally. The wastewater transport system is the link between the collection and treatment of the wastewater. The collection system includes, among others, the gravity flow sewage system from the house (or consumer) and service connection through street and main sewers up to the suction basins. The transport system, for which this manual was written, includes the suction basin, the sewage pumping station and the pressure pipelines.

In the Netherlands, municipalities and district water boards are the organisations responsible for the wastewater transport systems, as shown in Table 1.1.

Table 1.1 Information about pumping stations and pressure pipelines in the Netherlands (*Source:* Sewage statistics 2009–2010, RIONED Foundation).

Organisation	Capacity of End Pumping Stations [m^3/h]	Pressure Pipelines (Excluding Pressure Sewer) [km]
Municipality	960,000	5500
District water board	1,067,000	7300

This is a supplement to the existing and generally accessible information such as the Dutch Sewage Guideline. Modules B2000, B2100 and B2200 focus on the hydraulic design of the gravity flow system and provide designed flow rates for sewage pumping stations. Module C6000, pumping station management, mostly focuses on the design of the pumping station, as well as on management and maintenance organisation. This manual is also a supplement to existing standards, especially the Dutch standards NEN-EN 752:2008 *Drain and sewer systems outside buildings* – mostly about gravity flow sewerage – and NEN-EN 1671: *Pressurized sewerage systems outside buildings* – about pressure sewerage. This manual completely focuses on the hydraulic aspects of the design and management of wastewater transport systems.

In addition, many organisations have manuals about the design and management of pumping stations. These manuals mostly describe civil engineering, mechanical and electro-technical issues.

For now, this manual 'Hydraulic design and management of wastewater transport systems' should not be viewed as a replacement of such manuals, but as a supplement.

The following stages are recognised in the life cycle of a pipeline system (see also the Dutch standard NEN-EN 3650 'Requirements for Pipeline Systems'):

- Design
- Construction and testing
- Usage stage (operational management)

Before the design, the development stage takes place, also known as the preliminary design. The preliminary design is mostly determined by the usage requirements (functional requirements) and planning aspects. The design stage can be divided into the basic design stage and the detailed design stage. In the basic design stage, the definite points of departure (schedule of requirements) for the design are determined. In the detailed design stage, the calculations, drawings and specifications are established for the realisation and operational management stage. There is no fine distinction between the two design stages and, in this manual, it is summarised as 'design'.

The flow chart in Figure 1.1 describes the scope and interconnectivity of this manual. The starting point is that the preliminary design is available, although some points of attention are still mentioned. Therefore, this flow-chart emphasises the design of the transport system, followed by a chapter about the delivery of the installation, which describes how to test whether the built installation complies with the hydraulic design criteria. The construction of the installation is a stage that takes place between the design stage and delivery. In this stage, there are no specific hydraulic focal points and, therefore, this manual does not include a separate chapter about the building stage of the transport system.

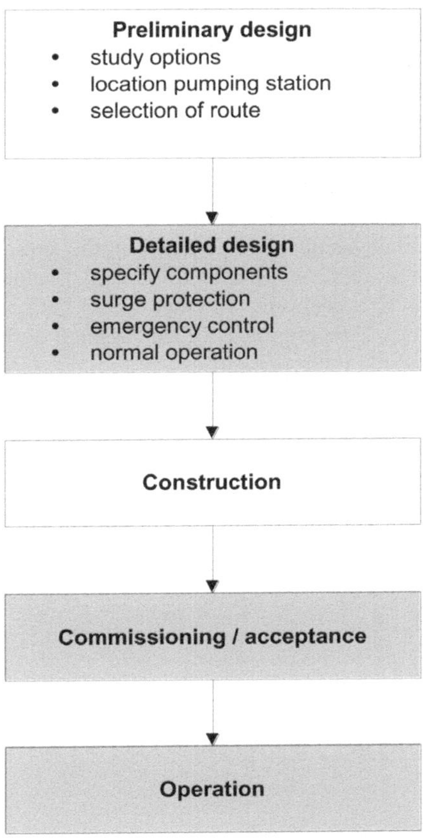

Figure 1.1 The stages of a wastewater pressure main. The highlighted topics are covered in this manual.

During the utilisation stage, the purpose of this manual is to maintain the desired capacity at minimal societal costs.

1.3 AUTHORS AND EDITORIAL STAFF

In 2010, the first version of this manual was drafted by Michiel Tukker (B.Eng), Kees Kooij (B.Eng) and Ivo Pothof (PhD) of Deltares. The editorial staff included:

François Clemens	Deltares/TU Delft (prof.)
Rinie van den Anker	District Water Board Rivierenland
John Driessen	Grontmij
Michiel Geise	Formerly: ITT Water & Wastewater
Jan Kranendonk	Rotterdam Public Works
Christof Lubbers	Royal HaskoningDHV
Piet van Rosmalen	District water board Delfland
Frank van Zijl	District Water Board Brabantse Delta

In 2012, a second Dutch version was published in which the experiences and comments of users were included. This manual was translated into English and published in 2016 (this edition).

1.4 READER'S GUIDE

The connection between the various components of the piping systems and the mutual interaction means that the reader cannot read this manual sequentially in one go, but will often have to return to previous sections. The theoretical background information has been compiled in a separate annex. The emphasis of this manual is on describing the design and management processes, and the interconnectivity.

Chapters 2–6 focus on the design process, which is tested again in Chapter 7. Chapter 8 discusses the delivery and acceptance stage. Maintaining the hydraulic capacity is discussed in Chapter 9.

Chapter 2
Designing wastewater transportation systems

The flow chart of the design process (Figure 2.1) shows that the design of a wastewater transport system is an iterative process. Every choice influences other components, which is why choices made might have to be repeated.

This does not only apply for the design process; Renovations and adjustments are also included in the same iterative process, because every change made to the system may have a far-reaching impact on the other components. This is why a system adjustment cannot be seen as a separate process, but as a design process with additional preconditions.

If the design process concerns an existing system, it is desirable to include the current situation of the system in the design process and not to just assume the starting points of the previous design. The design process must, therefore, be undertaken integrally and not divided into separate objects.

In many cases, it will not be possible to fully comply with all the main functions. In such cases, assessments will have to be made and priorities will have to be set. An assessment can be made by providing the designer with thorough theoretical knowledge. This manual contains the necessary hydraulic knowledge for the design and management assessments regarding wastewater pressure pipelines.

The transport system includes the entire system, starting from the receiving basin (suction basin), subsequently the pumping installation, the pressure pipeline and the endpoint (treatment plant or another receiving basin). The receiving basin and pumping installation together form the pumping station. This manual discusses both components separately.

Additionally, there are other arrangements that are necessary to guarantee the integrity of the system (water hammer surge protections) and to carry out management activities. First, the boundary conditions must be clear before discussing the design. The primary boundary condition is the discharge capacity in dry and wet weather conditions. In addition to transporting wastewater, the system must also be tested for the following requirements:

- Transportation of gasses (mixed-in air or chemically or bio-chemically formed gases)
- Transportation of solid elements (sediment, floating waste)
- Reliable and safe operation, also in extreme situations (water hammer)
- Low maintenance

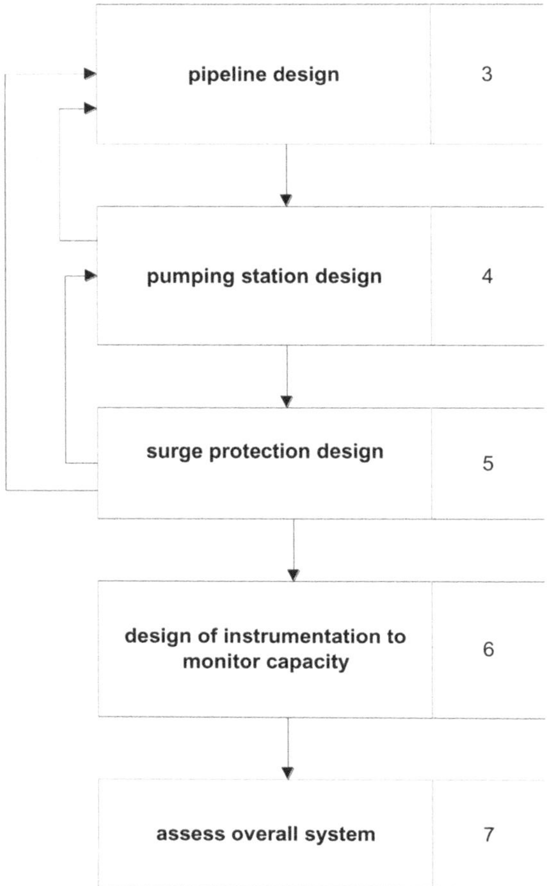

Figure 2.1 Flow chart of the detailed design process.

It is important to monitor the energy-saving and efficient operation of the system. Furthermore, the design must take into consideration the management stage, so that maintenance to the system can be carried out efficiently with minimum use of staff, as well as without delays and extra costs that can be avoided with a good design.

The following chapters will further elaborate on the design aspects of each component of the wastewater transport system.

Chapter 3
Pipeline design

We refer to a pressure pipeline when it is a pipe entirely filled with liquid (with the exception of local gas bubbles) at a positive pressure. The pressure is created in flat areas by a pumping installation. However, a main that connects a high reservoir with a low reservoir can also be considered a pressure main under the condition that no free surface flow occurs. Therefore, this is a pipeline flow driven by gravity.

In its simplest form, a pressure pipeline consists of a single pipe that connects a reservoir (suction basin) with another reservoir (basin, receiving construction, etc.).

A pressure pipeline can also connect to another pressure pipeline. Initially, this chapter will discuss the simple pipeline. In principle, the design aspects for a connecting pipeline do not deviate from this. The extra requirements can be converted into additional boundary conditions for the design process.

The pressure pipeline dominates the design and management process. The distance that the wastewater needs to bridge is often an established fact. In addition, carrying out a crossing (this includes all crossings such as culverts, horizontal directional drilling and dike crossings) with other objects (roads, railways, watercourses, dikes, etc.) is an important element in the question of capacity.

Most design choices regarding the pressure pipeline influence the design of the pumping station (pump and other appendages).

The design of a pressure pipeline is primarily determined by economic aspects, constructive aspects (choice of material, strength), and liquid-mechanical aspects. This manual only describes the liquid-mechanical aspects.

Figure 3.1 presents the preconditions and design activities for the design process of the pipeline(s).

3.1 BOUNDARY CONDITIONS

3.1.1 Flow rates

The flow rate is not a design choice, but is a fixed boundary condition. The pipeline system must be dimensioned to carry the maximum designed flow rate Q_{max}, usually the maximum expected flow rate in wet weather conditions. Most of the time (approximately 80%) we deal with dry weather supply, which means discontinuous pumping operation.

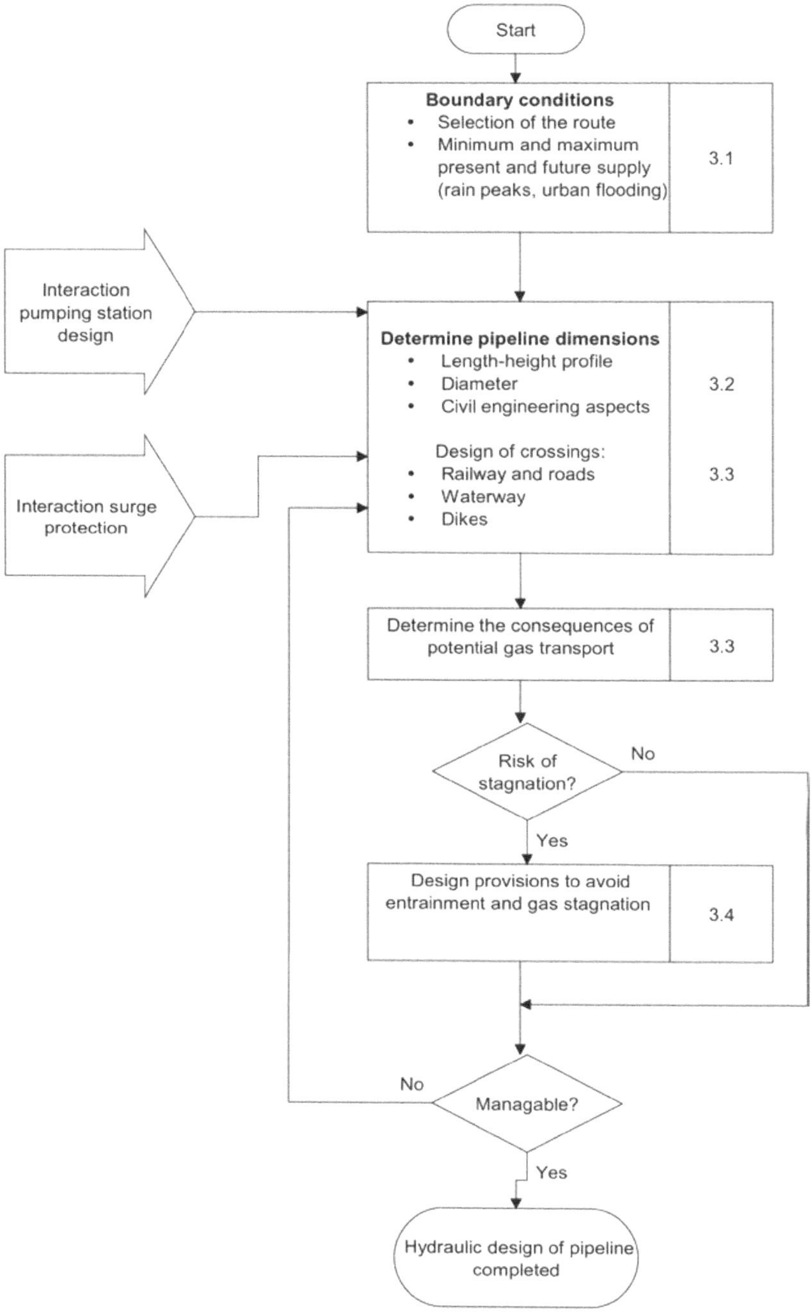

Figure 3.1 Flow chart for designing pressure pipelines.

Wastewater transport systems are driven by supply. The distribution between dry weather supply and wet weather supply, the installed pumping capacity and the type of control determine the distribution in the discharge. A pumping station with one on/off pump (while disregarding the variations in water levels in the basin) will always pump at the same flow rate regardless of the dry or wet weather supply. Only the duration that this flow rate flows through the pipeline depends on the supply.

3.1.2 Choice of route

The length of the pipeline is determined by the beginning location and the end location. Usually, the shortest distance possible is chosen while taking into consideration the existing infrastructure and zoning of the area, which is why the length of the pipeline is fixed.

During the preliminary design stage, it may be useful to investigate possible problematic points caused by the profile of the pipeline. When designing a wastewater pressure pipeline, the starting point is that the wastewater is in a situation of positive pressure in normal operation conditions, but also during standstill of the pump. At high points in the route, it may be that there is permanent negative pressure, which makes degasification possible. The released gas can cause extra loss of energy. Another source of energy loss can be the downward inclined pipe components of (drilled) pipes if gas bubbles accumulate here. That is why it is recommended to minimise the number of culverts and drilled pipes when choosing the route.

3.2 DETERMINING THE DIMENSIONS OF THE MAIN

The dimensions of the pipeline and the flow rate determine the loss of energy during operational circumstances. There are various equations for calculating this frictional loss (Darcy-Weisbach, Chézy, Manning). In the Netherlands, but also internationally, Darcy-Weisbach is used for entirely filled pressure pipes:

$$\Delta H = \frac{\lambda L}{D}\frac{v^2}{2g} \tag{3.1}$$

In which,

ΔH = Friction loss	[m]
λ = Friction coefficient	[-]
L = Pipe length	[m]
D = Diameter	[m]
v = Flow velocity	[m/s]
g = Gravitational acceleration	[m/s^2]

The friction coefficient λ is determined, among others, by the wall roughness of the pipeline, expressed in *k* (mm) – see Annex A.4.

Due to the planning boundary conditions, the pipe length is a parameter that can hardly be influenced. That is why the designer makes a great impact in determining the behaviour of the system when he chooses the diameter and, thus, the velocity in the pipeline. The choice regarding the material of the pipeline hardly has any influence on the capacity calculation, but does determine, to a large extent, the dynamic behaviour of the system (water hammer).

3.2.1 Profile of the pipeline

The height of the pressure pipeline is primarily determined by the local surface level. To limit earth moving, the pipeline will not be laid lower than what the local applicable regulations prescribe (frost proof, landowner, safe protection against damage due to digging, laying method, crossing objects, etc.).

An assumption is also that during all operation conditions, thus also during standstill of the pump, there is positive pressure in the pipeline. During negative pressure situations, as shown in Figure 3.2, there is a risk of creating gas bubbles in the system as a result of air entry from outside (leaking connections) on the one hand, and degasification of the wastewater (see Annex A.8.5) on the other hand. Primarily, the pipeline and connections must be resistant to all loads: external loads, such as ground and traffic loads, as well as internal loads, such as stationary and dynamic pressures.

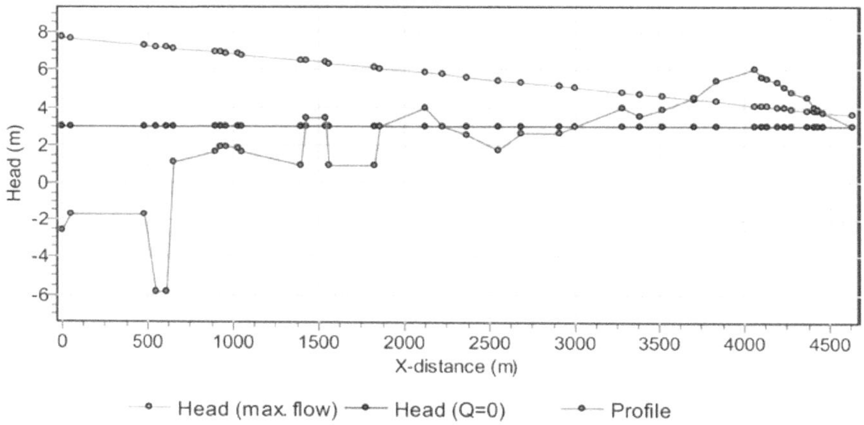

Figure 3.2 Hydraulic gradient and profile of the pipeline.

In order to avoid the risk of creating gas bubbles during negative pressures, the pipeline must be laid deeper or the discharge level must be increased. Laying the pipeline deeper results in higher costs during construction and might bring-in permit-specific requirements (for example in dike crossings). In the latter case, it may be sensible to lead the pressure pipeline directly into the elevated inflow of the treatment installation, if the pressure pipeline pumps the wastewater into a treatment installation. This way, negative pressures can be prevented in sections along the pipe route. However, this does increase the static head of the pump, yet the additional pumping station at treatment plant on the treatment installation becomes unnecessary, which is why the total energy consumption of the system will not increase.

If the pressure pipeline leads into a basin, installing an elevated weir might be considered. However, two disadvantages should be taken into account: 1) The energy consumption will increase structurally and 2) the overflowing water will create a stench problem because of released hydrogen sulphide.

In hilly areas it might be desirable to avoid a high point in the route and, therefore, to divert the pipeline. In this case, the length of the pipeline increases and, therefore, also the friction loss.

Which option is more desirable depends on the consequences of the various options. Possible consequences are:

- Limited choices regarding the pump and purchase costs of the pump(s)
- Change in energy consumption
- Installation costs of the pipeline
- Changes in descending pipelines with the risk of gas accumulations

Descending pipeline components should be given extra attention. Gas is discharged slower in these pipeline components than in horizontal or ascending pipeline components.

A gas bubble can lead to substantial extra losses, especially in culverts and drilled pipes that reach a significant depth: the maximum is equal to the difference in height of the crossing (see Section 3.3).

However, pipelines that follow the surface level under a slightly descending angle can also be a source of gas bubble stagnation. Stagnation of gas bubbles occurs when the pipeline is steeper than the hydraulic gradient. Assuming a friction loss of between one and five metres per 1000 m pipeline, a descending slope of the pipeline with an angle of 0.1 up to 0.3 degrees can already lead to problems. The gas bubble is entrained as long as the hydraulic gradient is steeper than the slope of the pipeline, although the speed of the gas bubble is slower than that of the water velocity. The balance of forces on the gas bubble causes the bubble to go to a point where there is lower pressure. If the hydraulic gradient is less than the gradient of the pipeline, the gas transportation will depend on the flow velocity in the pipeline (see Annex A.8).

The transport capacity of gas is very small in descending pipelines of culverts and drilled pipes. It takes hours to degrade a gas bubble through pipe flow. Therefore, over time a pipeline might be filled with gas over the entire descending length. Figures 3.3 and 3.4 show both situations.

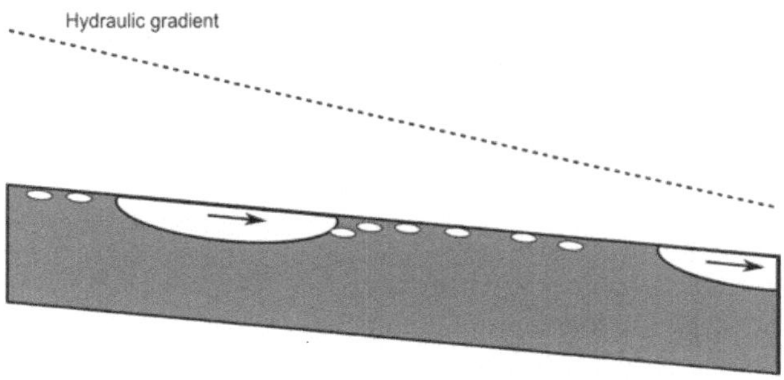

Figure 3.3 Behaviour of air bubbles when slope is less than the hydraulic gradient.

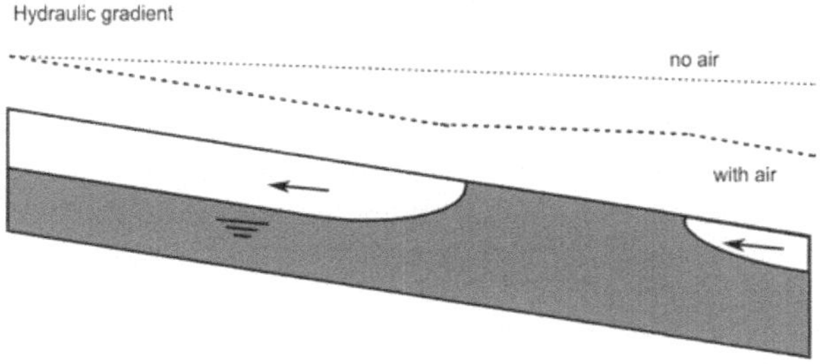

Figure 3.4 Behaviour of air bubbles, if the slope of the pipeline is steeper than the hydraulic gradient, and the velocity of the fluid flow in area 1 or 2 (see Figure 3.8).

If the air bubble continues to expand up to the bottom of a descending pipeline, then there is an additional loss of energy that equals the total height of the descending pipeline minus the resistance loss of the entirely filled pipeline.

$$\Delta H_{gas,\max} = L \sin\theta - \Delta H_{fric} \tag{3.2}$$

In which,

$\Delta H_{gas,\max}$ = Maximum energy loss of the gas bubble	[m]
L = Length of descending pipeline	[m]
θ = Slope angle of descending pipeline	[°]
ΔH_{fric} = Friction loss over descending pipeline without gas bubbles [according to the equation 2.1]	[m]

The designer must also realise that a pipeline, which is put into operation, must also be made air-free properly. Annex A.8 provides the theoretical relationship between the necessary velocity of the fluid flow and slope angle.

An ideal profile that does not allow any gas pockets lies flat or in an ascending slope. If a descending pipeline is necessary, then the preference is for steep slope angles (>60°, in which 90° slopes realise the highest gas transportation). Research has shown that gas can be discharged faster with steep slope angles (see Section 3.3).

3.2.2 Diameter

The internal diameter is decisive for the loss of energy. As the length of the pipeline and the maximum flow rate are usually fixed, only the internal diameter remains as the most important design parameter. The necessary wall thickness, and thus the external diameter, is determined by the pipeline strength requirements.

Advantages of choosing a smaller diameter are:

- Cheaper to purchase (less material)
- Cheaper to lay (less volume of earthwork)
- Shorter residence time in the pressure pipeline
- Improved discharge of gas bubbles

The disadvantages of a smaller diameter are:

- Higher loss of energy: a 10% smaller diameter results in 60% more loss of energy (the diameter appears to the power of five in the friction equation).
- Shorter utilisation period: future increase of capacity will lead to replacement more quickly due to occurring energy losses.

The first two advantages influence the construction costs of the system. Both disadvantages bear consequences for the operating costs. It is rarely favourable to choose a smaller diameter, because in practice the operating costs of a wastewater transport system are decisive.

A smaller diameter has a positive influence on the residence time and the release of hydrogen sulphide (H_2S). The higher energy costs of a smaller pipeline diameter usually outweigh the investing in measures regarding the discharge point.

The maximum speed used in the Dutch wastewater practice is between 1.0 and 1.5 m/s (2.0 m/s in pipelines with a diameter larger than 1.5 m) with up to 50 meter pump head. Over the lifespan, the energy costs are usually too high with higher heads.

The minimum necessary speed in the pipelines depends on two aspects:

- Pollution/sedimentation
- Creation of gas bubbles

Sediment transport in pipelines depends on the particle size of the sediment. This particle size is unpredictable in wastewater systems, which is why it is impossible to define a scientifically based minimum speed.

Experimental research (Fair, 1968) has shown that with a speed of 0.6 or 0.7 m/s, a pressure pipeline has a self-cleaning capacity due to intermittent operation. In practice, a lower speed of 0.5 m/s is usually used as minimum transport velocity which is comparable to the values found by Fair (Fair 1968).

The Flemish code (KU Leuven, 2004) uses minimum shear stresses of 2 N/m^2 in dry weather supply pressure pipelines of separate systems and 3 N/m^2 in pressure pipelines with mixed supply or wet weather supply. The increased shear stress in pressure pipelines of mixed systems and wet weather supply pipelines is desirable due to the transport of sand particles. These shear stresses should be achieved at least once a day and can, therefore, be considered as minimum design speeds for dry weather conditions (Figure 3.5) and Table 3.1.

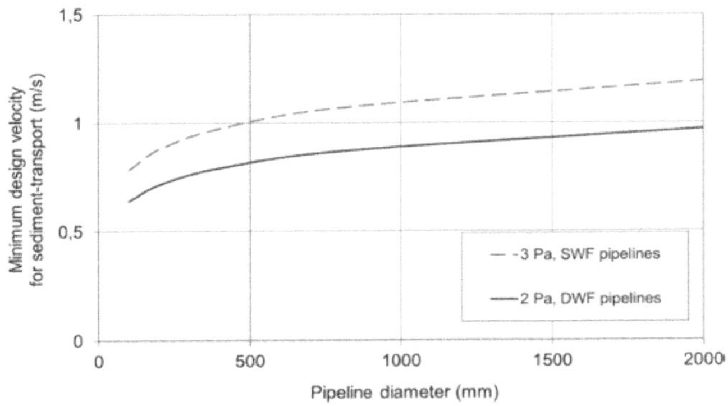

Figure 3.5 Recommended minimum designed speed in pressure pipelines during dry weather, based on 1 mm wall roughness. In case of lower designed roughness, the designed speeds increase in order to achieve the same shear stress.

Table 3.1 Summary of minimum recommended transport speeds in wastewater pressure pipelines.

Situation	Speed [m/s]	Comments
Minimum transport speed	0.6–0.7	According to Fair (1968)
Speed that should be reached at least once a day during dry weather conditions	0.64–0.97	According to the dry weather curve in Figure 3.5
Speed that should be reached at least once a day during wet weather conditions in pressure pipelines of mixed systems	0.8–1.2	According to the wet weather curve in Figure 3.5

3.2.3 Wall roughness

In principle, the equivalent wall roughness value *k* is a characteristic of the pipe material. The roughness contribution of pipe welding, sockets or connections is often also discounted in the roughness equivalent, which is why the roughness value *k* has become more of a calculation unit. The literature offers tables in which this *k*-value is mentioned per pipe material (see Annex A.3). However, these values apply to clean pipes.

With wastewater, after some time, a fouling layer develops on the internal wall and, as a result, the original pipe material no longer has any effect. These fouling layers (film) are quite thin (in sizes of 1–5 millimetres) and only have minor influence on the internal diameter of the pipe. This fouling layer can even have a positive effect if the original pipe wall is rough. Other processes that influence the wall roughness are sedimentation and scaling (for further reading, see management aspects and theory).

The degree of roughness strongly depends on the quality of the wastewater. That is why, in practice, the *k*-value from the table in Annex A.3 is not used.

Generally, when designing wastewater transport pipelines, a wall roughness, *k,* of 0.25–1 mm is used.

It is important to clarify the effect of these variations in *k*-values. This effect is best illustrated when the system or pipeline characteristics are calculated.

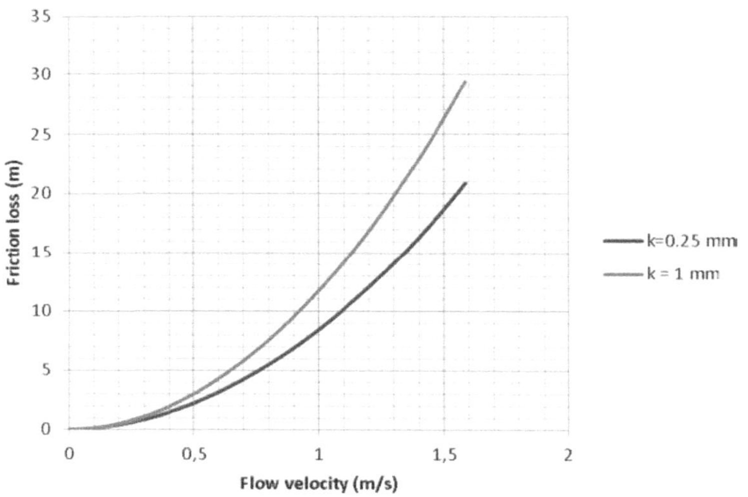

Figure 3.6 Effect of wall roughness on the system characteristic for pipeline with D = 250 mm and length of 2 km.

The influence of the *k*-value decreases as the diameter increases, because the relative wall roughness (the wall roughness divided by the pipe diameter, *k/D*) is used.

3.2.4 Pipeline material

A designer can choose from various materials for a pipeline system. Aspects that influence his choice are strength, costs, lifespan and manageability during construction. Sustainability of the material seems to also becoming one of these aspects, although it is not yet clear how best to inter-compare material sustainability.

Hydraulic aspects are almost never a reason for choosing a specific type of material. The various *k*-values of the new materials no longer seem to make a difference due to the fouling layer (biofilm) created during utilisation. Furthermore, for all materials that are used, there are strength classes for the pressures that are common in the Netherlands. The same applies for the quality of the wastewater. All pipeline materials are resistant to the wastewater that is common in the Netherlands.

The maximum positive and negative pressures follow from the water hammer analysis. Together with the other loads (ground load, traffic load, etc.), the strength analysis will show whether the pipe is able to cope with the expected positive and negative pressures. The Dutch standard NEN 3650 describes the strength analyses. For example, thin-walled pipes are sensitive for implosion (folds or buckling) at negative pressure (see also Annex A.7.4).

Until the early 1970s, asbestos cement was a popular material, because it was cheap and light, and, therefore, easy to manage during construction. In the 70s and 80s, the danger of asbestos cement was recognised and it was no longer used. A significant proportion of wastewater pressure pipelines in the Netherlands are still asbestos cement pipes.

The same advantages that applied to asbestos cement, namely that it was cheap, light and easy to manage, now apply to the pipe material PVC. In normal applications, this is the most commonly used material for most of the diameters in the range from 200 to 400 mm. The long lifespan, as recent research has shown, is an additional advantage.

Polyethylene (PE) is usually used in horizontal directional drilling, though bores made from cast iron are also common. The remaining pipe materials, pre-stressed concrete, cast iron, glass-fibre reinforced plastic (GRP or GFRP) and steel, are typically employed for relatively large diameters and special conditions.

3.3 CROSSINGS

The behaviour of gas bubbles in pressure pipelines is unstable and strongly depends on the slope angle of the descending pipe. The CAPWAT research (2003–2010) provided much knowledge about this. When gas bubbles stagnate and accumulate in the pipeline, this can lead to significant local loss of energy that is practically equal to the level of the gas volume (see Figure 3.7).

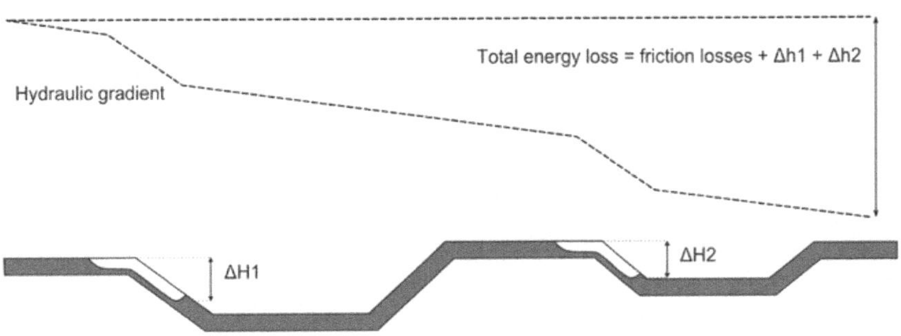

Figure 3.7 Loss of energy due to gas bubbles visible in the hydraulic profile.

Stagnation only occurs in descending pipe components. This aspect plays a role when an object (e.g. road, watercourse, railway, dike) needs to be crossed. However, it can also occur in mildly descended pipelines if the slope angle is larger than the hydraulic gradient. Due to the complex character, a separate section has been dedicated to this issue (see Section Figure 3.3).

A minimum speed is required in order to initiate gas transport in a descending pipeline under slope angle θ. If the speed is less than this minimum speed, then the gas will accumulate in the pipeline until the bubble reaches the bottom of the descending pipeline. In such a case, the maximum additional loss of energy needs to be taken into account, which is equal to the height difference over the descending pipeline. However, the necessary flow speed also strongly depends on the pipeline diameter. That is why we use a dimensionless flow speed - the flow number:

$$F = \frac{v}{\sqrt{gD}} \tag{3.3}$$

In which:

F = Flow number	[-]
v = Flow speed	[m/s]
g = Gravitational acceleration	[m/s^2]
D = Diameter	[m]

The flow number allows for phenomena that are dominated by gravity to be transferred to other pipeline diameters. The free surface flow under a gas bubble with a hydraulic jump is an example of this. The flow number does not take into account the transfer effects with regard to the water quality, but research has shown that these transfer effects are negligible in pipeline diameters larger than 150 mm (Tukker, 2007).

Figure 3.8 shows the connection between the slope angle θ of the pipeline and the necessary flow number for removal of gas. These flow numbers derive from a calculation model that was developed during the CAPWAT research (2003–2010) (Pothof, 2010).

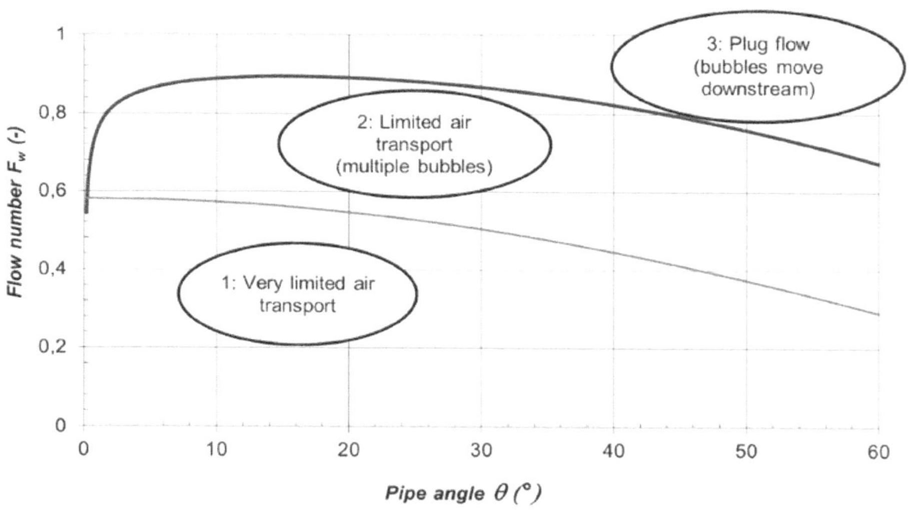

Figure 3.8 The necessary flow number for the removal of gas; required flow number for air transport depending on the angle of the downward sloping pipeline.

Figure 3.8 is divided into three areas. In area one, there is little gas transport. The flow speed in the pipeline is too low to transport gas bubbles to the lowest point. Gas is mostly discharged as a dissolved matter in the liquid, which results in a low gas discharge. As a result, the entire descending part of the pipeline might be filled with gas, which leads to a loss of energy that equals the height difference over the descending pipeline.

Area two is located between the criteria for gas **bubble** transport and gas **volume** transport. In this area, the gas volume is divided into several separate gas bubbles that are spread over the length of the descending pipeline. Small bubbles are carried with the liquid flow and transported past the lowest point. The loss of energy in continuous gas supply significantly decreases (the largest reduction in loss of energy occurs in this area) the more the flow number increases.

Area three is the area above the criteria for gas volume transport. In this area, gas volumes are transported, regardless of their size. The loss of energy caused by the gas volumes is negligible (depending on the amount of discharged gas).

The aforementioned relationship (Figure 3.8) between slope angles and the necessary flow number was validated during the CAPWAT research for angles between 5° and 30°and this applies to all pipeline diameters and lengths (Pothof, 2010).

Figure 3.8 cannot be used for angles larger than 60°. Other physical processes play a dominant role in these slope angles. Research (Lubbers, 2007) has shown that the necessary flow number for gas volume transport in a vertical pipeline (slope angle is 90°) is 0.4 (F_{90} = 0.4) and, thus, significantly smaller than the necessary flow number for 60° (see Figure 3.8). Gas bubbles, therefore, are most efficiently transported downwards in a vertical pipeline.

The pipeline speed is shown below as a function of the diameter for several typical flow numbers ($F_w = 0.4$; $F_w = 0.6$ and $F_w = 0.9$).

A comparison of Figure 3.9 and Figure 3.5 shows that the necessary velocities for gas transport are significantly higher than for sediment transport, especially in pipelines with a diameter larger than 200 mm. The following section discusses several solutions for management of gas accumulations in pressure pipelines.

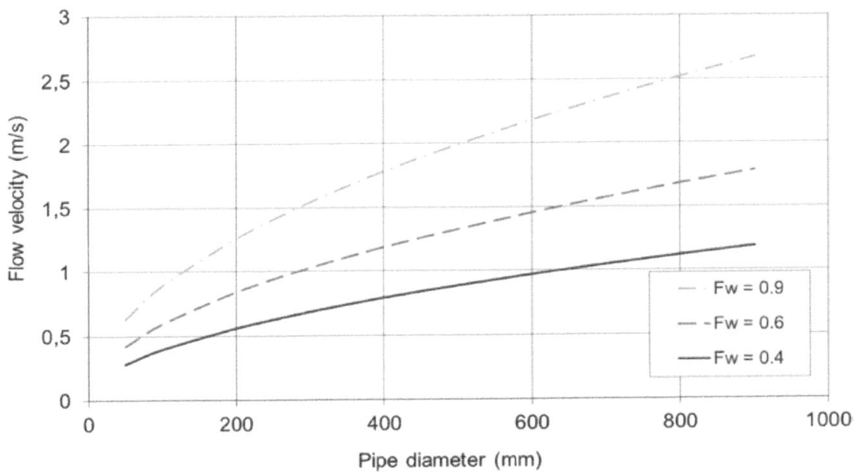

Figure 3.9 Necessary pipeline speed to reach a certain flow number as a function of the pipeline diameter.

3.4 ARRANGEMENTS FOR THE PREVENTION AND REMOVAL OF GAS BUBBLES

Usually, wastewater transport systems are designed with a maximum flow speed of 1.0 or 1.5 m/s to keep the energy consumption low. The consumed capacity is proportional to the flow speed to the power of three ($P \propto v^3$). Designing for complete gas transport (flow number is at least 0.9) is often not desirable from an energy point of view. Additionally, it is unrealistic for pipelines with a large diameter because too high speeds are necessary (Figure 3.9). Therefore, other measures must be taken for this situation.

An important motto regarding air pockets is 'prevention is better than cure'. In practice, the cause of gas bubbles is mostly determined by air intake at the pumping station. See chapter 4 for prevention/limiting this. The solutions below concern the pressure pipeline.

3.4.1 Reduce the diameter locally

In order to improve gas transport in the descending parts of the pipeline, the crossing can be carried out with a smaller diameter. That way, the flow speed in the descending parts of the pipeline is higher, which has a positive effect on the gas transport.

Equation (3.4) provides the maximum diameter as a function of the water flow rate and a desirable flow number.

$$D = \left(\frac{4Q}{\pi F_w \sqrt{g}} \right)^{\frac{2}{5}} \tag{3.4}$$

In which:

D = Pipeline diameter	[m]
Q = Discharge	[m^3/s]
F_w = Desirable flow number	[-]
g = Gravitational acceleration	[m/s^2]

The resulting diameter is the maximum diameter for the crossing in which the necessary flow number can still be achieved. The necessary minimum flow number is provided by the gas volume transport criteria (see Figure 3.8). An advantage of this approach is simplicity and reliability. Due to the high flow number, the gas bubble will never remain in the crossing.

A disadvantage is that the pipeline diameter greatly influences ($\Delta H \propto \frac{1}{D^5}$) the pump head and, therefore, also the consumed capacity (see Annex A.5). As a result, the required capacity might increase significantly. Because all the crossings are carried out with a smaller diameter, the extra frictional losses are usually small. It is very important to consider the friction losses and the risk of problematic gas bubbles in the design stage. The smaller pipeline diameter causes permanent additional loss in energy, in contrast to the gas bubble, which causes a risk for more additional loss in energy. If sufficient attention is paid to the prevention of air intake during the design or redesign of the pumping station (see Section 4.5.2) then a smaller diameter does not need to be used in crossings.

A strongly varying pipeline diameter can also cause problems with pigging of the pressure pipeline. The allowed diameter reduction strongly depends on the type of pig. A soft pig can handle a diameter reduction of 30% and sometimes up to 50%, depending on the material of the foam pig. A hard pig may not allow for diameter reduction. The risk of a pig getting stuck in a pipe must be prevented at all costs. The choice of pipeline diameter variations in relation to pigs must be given sufficient thought in the design process.

3.4.2 Air-release valve

If reducing the diameter does not provide a good solution (too many problems for cleaning activities, too high friction losses) then an air-release valve must be applied to the pressure pipeline upstream of the crossing.

Air-release valves on the pressure pipeline are a sensitive subject among the managers and operators of wastewater transport systems, because the reliability in practice is not sufficient because of grease and floating waste. This pollution can prevent the air-release valve from opening and closing or can clog up a connecting pipe, a manhole or a branch pipe. So far, not enough research has been done into the most favourable configuration of an air valve, which catches air and simultaneously does not clog up quickly with floating waste. That is why no clear recommendation regarding this issue can be given.

Because of stench nuisance, attention must be paid to the accessibility and environmental factors when designing an air valve. Initially, a manual air valve may suffice, which is a requirement even before filling the system. If there is no gas stagnation (because sufficient preventive measures have been taken at the pumping station) then an automatic air valve is not necessary. When designing the air valve location, one should keep in mind the necessary overall dimensions of the air valve and stench filters that are available on the market.

Because positive pressure is necessary in the pipeline to discharge gas, attention must be paid to the hydraulic gradient (see Section 3.2) when designing the air valve. The placement of the air valve can strongly influence the effectiveness. Air valves are usually placed on the highest point of the pipeline because of the intermittent effect of wastewater systems. There is not always an obvious high point, the most suitable location is not always accessible or it might be too far from the crossing at which gas can accumulate.

Therefore, a simple solution for wastewater with limited floating waste is placing a branch pipe with an air valve right before the descending section of the crossing (Figure 3.10). The minimum size of the lateral branch of the branch pipe is determined by the diameter of the main. Research on clean water has shown that the minimum width of the lateral branch must be 0.5 times the diameter of the main pipe (Wickenhäuser, 2008). With this size, all gas bubbles that pass through the top of the supplying pipeline are caught in the branch pipe. The minimum height of the branch pipe should be 0.3 times the main pipe diameter. If the height of the branch pipe is smaller, the gas bubbles can escape from the branch pipe and can be carried with the flow (see Annex A.8).

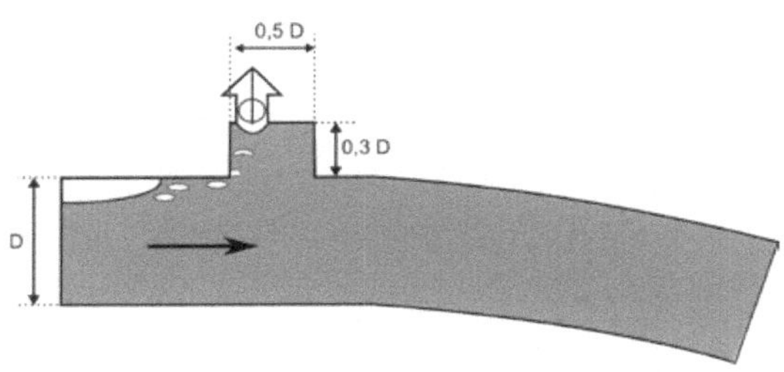

Figure 3.10 The air valve is mounted on top of the branch pipe.

3.5 NETWORK SYSTEMS

In practice, several pumping stations are often connected to a main pipeline that flows to the sewage water treatment plants. As a result, pumping stations influence each other. This should be taken into consideration in the design.

This also means that modifications made to a pumping station influence all the other pumping stations in the system. A capacity upgrade of one pumping station can lead to a capacity problem for another pumping station due to the increased pressure on the location where the pipes meet. Essentially, the design process of a network does not differ from that of a simple system.

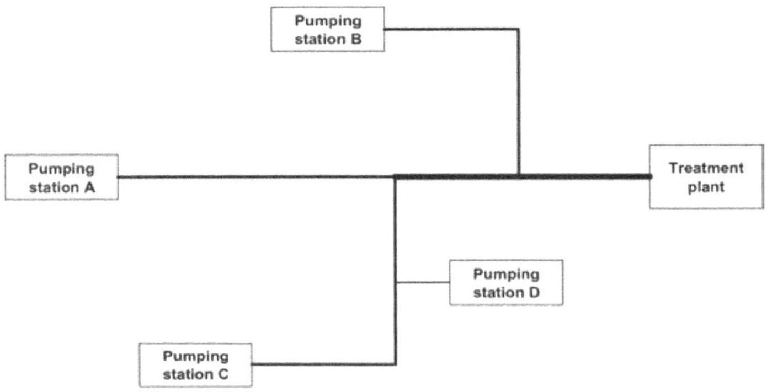

Figure 3.11 Example of a wastewater transport network.

Currently, each pumping station in the system has various pump heads depending on whether the other pumping stations in the system are operating. The various operating points are clarified by means of a combined pump/system characteristic graph (see Figure 3.12).

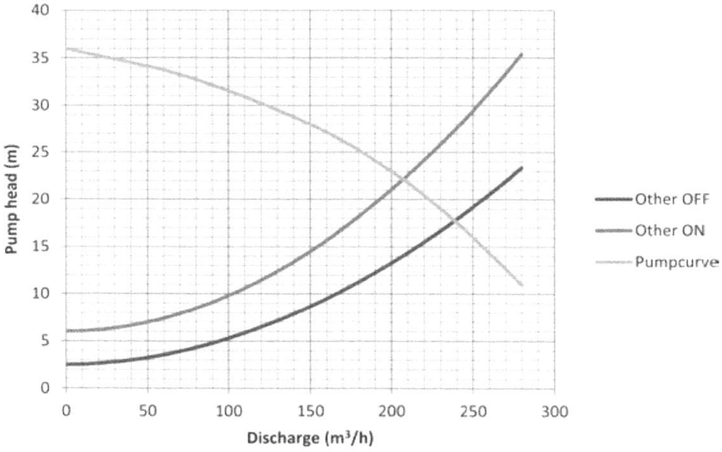

Figure 3.12 Operating points of the pump in different situations.

The designer should not confine himself only to the new or to-be-modified pumping station, but should also check the operating points of the existing, unchanged pumping stations.

Chapter 4
Designing the pumping station

The pumping station is the driving force of the wastewater transport system. The pumping station consists of a pumping unit and a receiving basin. The receiving basin is actually part of the pumping station and cannot be considered a component separate of the pump unit.

There are various types of pumping stations, each with specific points of focus:

Type	Description
Sewage pumping station	The receiving basin is directly connected to the sewage system. The receiving basin can consist of an open basin or an underground sewage stretch. The supply varies relatively slowly.
Intermediate pumping station	The usually smaller receiving basin receives wastewater from other pressure pipelines and sometimes also from its own sewage system. Controls must be adjusted to the size of the receiving basin and at least as fast as the controls of the supplying pumping stations.
Booster pumping station	A booster pumping station does not have its own basin. Control of the supplying pumping stations must be adjusted to the control of the booster pumping station.

Figure 4.1 shows the boundary conditions and design activities for the design process of the pumping station that will be discussed in this chapter. The various sections provide the reader with an overview of the aspects that play a role in the design.

4.1 BOUNDARY CONDITIONS

4.1.1 Required capacity

The capacity of the pumping unit is a strict requirement. The required capacity can be divided into dry and wet weather supply capacity. In the Netherlands, dry weather supply capacity occurs approximately 80% of the time. Wet weather supply capacity is incidental, but is never the less decisive for the system.

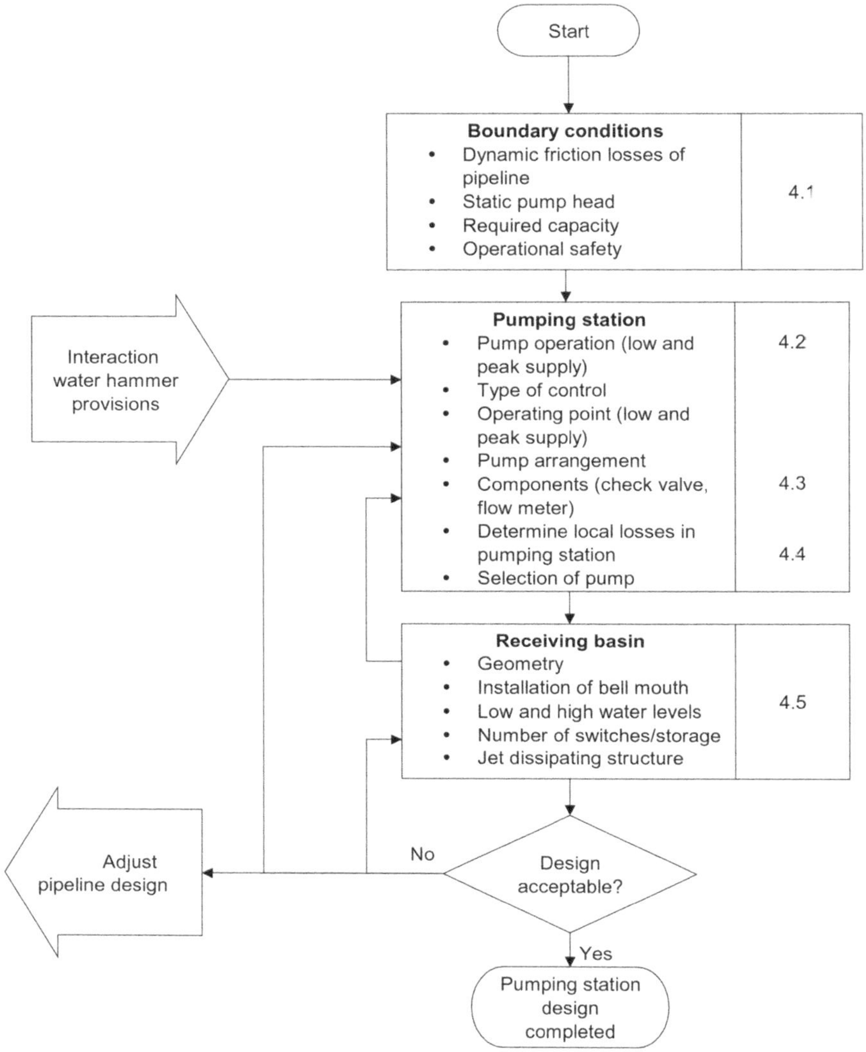

Figure 4.1 Flow chart of the design of the pumping station.

4.1.2 Pump head

The head of the pumping station consists of two components: the static and the dynamic head of a pump.

In pipelines to which only one pumping station is connected, the static head of a pump is the same as the height difference between the water level in the pumping basin and the water level at end of the pipeline (receiving basin or wastewater treatment plant).

Local losses do not play a role in an entirely filled pipeline (thus without gas bubbles in the system) and the dynamic head of a pump is determined by the friction loss in the pressure pipeline. The influence of bends is so small in relation to the uncertainty of the wall friction, that it is usually unnecessary to take into account the local losses in the pressure pipelines. However, local losses due to gas accumulation can have a major impact (more about this in Section 3.3). The local losses in the pumping station (including in check valves, shut-off valves) are decisive, because they can be significant due to the number of sharp bends and buckles, shut-off valves and check valves and the high flow speeds.

In complex networks, the pump head depends on the other pumping stations in the system, which creates a varying head and, therefore, a varying operating point for the pump. This should be taken into consideration when designing the pump installation and selecting the pump.

4.1.3 Operational safety

The operational safety of the pumping station is a requirement that is very hard to quantify and is mostly determined by the cost aspect. The costs of management and maintenance are often higher than the construction costs and are mostly determined by the design. Higher costs during construction of the system can pay off at the end in lower management and maintenance costs.

The percentage of time that pumps are available to pump can be used as an indicator of operational safety. However, this strongly depends on the conditions (geographic location, water quality, arbitrariness) and, therefore, cannot be used directly for comparing the systems with each other.

4.2 PUMP SET-UP

4.2.1 Wet or dry set-up

From a hydraulic point of view, there is virtually no difference between a wet and a dry pump set-up. However, there are consequences for the water levels at which the pump switches on or off. The choice between a wet and a dry pump set-up is often not a technical, but rather an organisational decision.

Many organisations have fixed criteria for this decision. A frequently used criterion is the weight of the pump. The capacity of the pump is also used as a criterion, but in practice this is the same as the weight of the pump (higher pump capacity means a heavier pump).

The pump in a wet pump set-up has to be lifted out of the pump chamber for maintenance and that is why many organisations establish requirements for the maximum weight of the wet set-up pump. The requirements are often the limits of the crane or the lifting equipment that the organisation has available.

Opinions vary strongly about the technical, historical and emotional considerations. In summary, the following considerations can play a role when choosing between a wet and a dry set-up:

- Organisation of the management
- Occupational Health and Safety aspects
- Weight per pump or capacity of the pumping station
- Installation costs of the pumping station
- Available space underground for building

4.2.2 Number of pumps

The number of pumps is determined by the distribution between the required wet and dry weather supply capacity. Nowadays, many sewage-pumping stations are equipped with two pumps; one of these pumps is a standby pump, so that the pumping station can always switch in case of failure of the other pumps.

Because the pumps must be selected based on their wet weather supply capacity, the pump will often have to switch to situations of dry weather supply capacity (depending on the difference between wet and dry weather). The switching frequency can be reduced by using a speed control and that way a single pump can reach higher flow rates. The effect of a speed control on the behaviour of the pump is discussed in Annex A.6.

4.2.3 Vertical or horizontal set-up

Dry pumps can be set up vertically or horizontally (this refers to the orientation of the motor shaft), usually in order to limit the necessary space that the pump occupies. However, this does have hydraulic consequences. The preference for a vertical or horizontal set-up of the pump depends on the choice between a wet and a dry set-up.

Dry set-up pumps with large dimensions are sometimes set up vertically to limit the required space. In the past, pumps were also set up vertically to create distance between the pump motor and the pump housing.

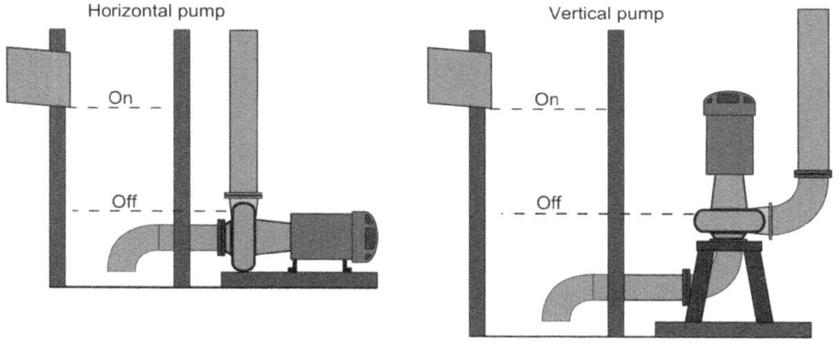

Figure 4.2 The difference between a horizontal and a vertical pump set-up.

In a dry situation, the horizontal pump set-up is favoured over the vertical set-up for the following reasons:

- Simpler pipework
- Lower friction losses (vertical pumps need two additional 90° bends at locations with high flow velocities)
- Water level at which the pump switches off must be higher than the horizontally set-up variant (or the chamber must be deeper).

Wet pumps are always placed vertically, because of the pump's attachment to the discharge connection. In addition, with a vertical set-up, the inlet of the pump is located on the lowest point, which is favourable for the water levels at which the pump switches on or off.

In the past, wet pumps were also placed horizontally so that the motor could be cooled longer by the water. This is not necessary with modern pumps/electric motors.

4.3 COMPONENTS

The dynamic losses in the pumping station usually vary between 0.5 m and 2 m and are primarily determined by the flow speed in the pipeline system. An important factor in the dynamic losses is the resistance in the check valve. It is important to choose a check valve that will be completely open during pump operation.

4.3.1 Shut-off valves

Various shut-off valves are needed in the pumping station in order to facilitate maintenance to the pumping unit. These are indicated in the figure below.

A dry pump needs at least two shut-off valves in order to allow maintenance to be carried out on the pump: one on the suction side and one on the downstream side of the check valve. One shut-off valve on the outlet side of the pump is enough for a wet pump (see Figure 4.3).

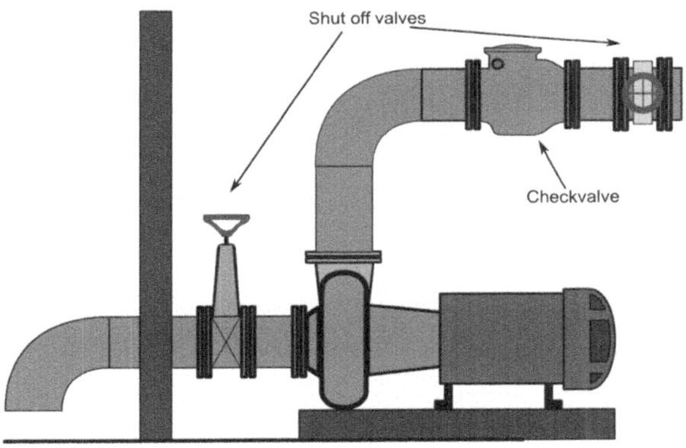

Figure 4.3 Location of the shut-off valves and the check valve.

In addition, a third shut-off valve should be placed at the beginning of the pressure pipeline (this applies to wet and dry pumping stations). This allows closing off the entire pumping station from the system. This is necessary in order to carry out maintenance to the pipe work in the pumping station, to measure capacity, and/or to carry out work on the pumping station.

Because the shut-off valve is used to allow maintenance, it must have complete free passage when open in order to prevent clogging. If the shut-off valve is (partly) clogged, it might no longer be possible for it to close entirely. In practice, gate valves are mostly used.

4.3.2 Check valve

A check valve is a relatively simple device that consists of a valve housing around one or several rotational or translational shut-off bodies with the purpose of allowing liquid to pass only in one direction. The movement of the shut-off bodies are primarily controlled by the liquid. In some types, this movement is also influenced by a damping mechanism, which slows down the movement of the shut-off bodies (usually, only during the last part of the shut-off phase).

The stationary characteristic, which gives the relationship between the pressure difference over the valve and the flow rate, describes the stationary behaviour of the check valves.

The pressure level difference over the valve is:

$$\Delta H = \xi \frac{v^2}{2g} \tag{4.1}$$

In which:

ΔH = Pressure level difference	[m]
ξ = Coefficient of resistance and losses	[-]
v = Flow speed	[m/s]
g = Gravitational acceleration	[m/s^2]

The dimensionless coefficient of resistance or losses, ξ, is a function of the valve position, θ. The stationary characteristic gives the relationship between the pressure level difference/pressure difference and liquid speed/flow rate. Figure 4.4, which plots the pressure level difference as a function of the liquid speed, shows the characteristics of a translational type of check valve (nozzle) for two different springs.

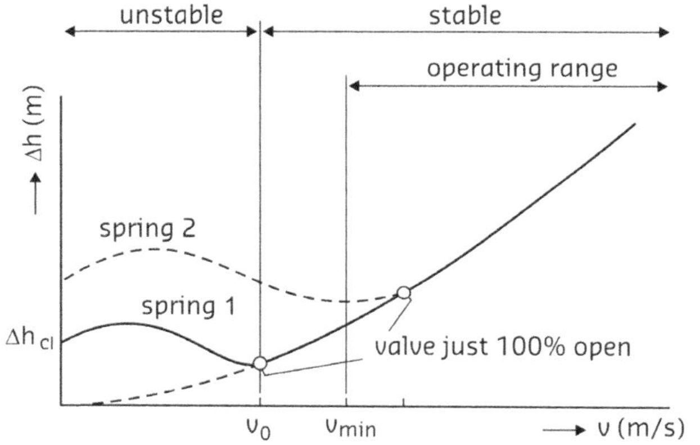

Figure 4.4 Stationary characteristic of a check valve.

The pressure difference, Δh_{cl}, is the value while the valve begins to open. The critical speed, v_o, is the flow velocity at which the valve is completely open. The characteristic is parabolic for liquid speeds larger than the critical speed, because ξ is constant for a completely opened valve. When the flow rate decreases, the check valve begins to close from the moment that the critical speed, v_o, undershoots and in which the pressure level difference over the valve might increase slightly due to an increase in the resistance.

Depending on the type of valve, the stationary characteristic may be influenced by springs (stronger, weaker), weights and balls (heavier, lighter) and the stroke or maximum rotation angle of the moving components. Section A.6.3 describes the dynamic behaviour of the check valve.

When designing the pump installation, attention must be paid to the fact that the orientation and installation of the check valve influences the stationary and dynamic characteristics of the valve.

Orientation means the flow direction of the valve. The orientation of the flow direction is horizontal in the check valve in Figure 4.3. The manufacturer of the valve must indicate what the proper installation requirements are for a certain valve.

4.4 DETERMINING THE OPERATING RANGE AND SELECTING THE PUMP

As soon as the static and dynamic pump heads of the system are known, the operating range of the pump can be determined.

The required pump installation's pump head, *H*, in the pumping station depends on:

- The static head of a pump (the pressure difference between the receiving basin and the downstream discharge construction)
- The dynamic head of a pump (pressure loss that is caused by the pressure pipeline)
- The pressure losses in the pipelines and appendages (shut-off valves, check valves)

The friction losses in the components can be determined based on literature values (bends, buckles, pipework) or are indicated in the specifications of the manufacturer (shut-off valves, check valves). The static and dynamic heads depend on the design of the pipelines. However, the pump head might vary in situations where more than one pumping station is connected to the pressure pipeline (see Figure 4.5).

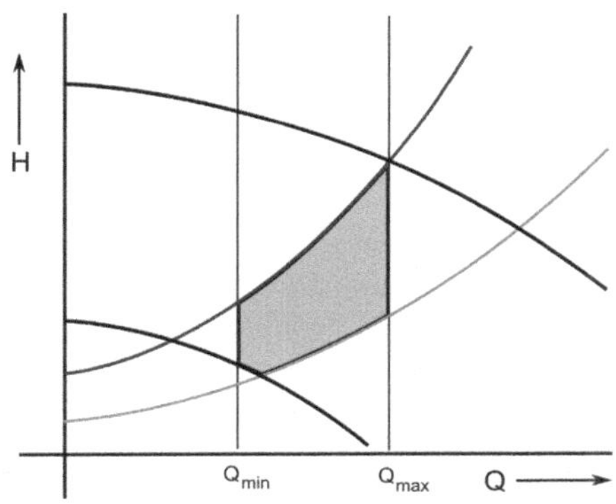

Figure 4.5 The operating range of the system (with varying pump heads and levels of wall roughness) and accompanying pump curves (variable speed pump).

Combining the pump characteristics and the system characteristics, and indicating the minimum and maximum flow rates in the system characteristics, can give a clear picture of the operating range (see Figure 4.5).

The actual pump choice can best be left to the pump manufacturer. With a clearly defined system characteristic and operating range, the manufacturer should be capable of choosing the right pump for the system.

4.5 RECEIVING BASIN

Various guidelines and manuals describe in detail the hydraulic design of pump chambers (see annexes for more information). This section provides a brief description of the various aspects that are involved in the hydraulic design of pump chambers and, where necessary, will refer to the literature.

A properly designed receiving basin complies with the following requirements (in order of priority):

1. The storage has optimal frequency control for the pumps and the pipeline system;
2. The pump does not suck in air;
3. The pump discharges sediment and floating waste; and
4. The approaching flow of the pump is optimal.

In practice, it is nearly impossible to design a receiving basin that complies with all the requirements, because the aforementioned requirements can be contradictory. This will be discussed further.

4.5.1 Storage of the receiving basin

Due to cost considerations, the receiving basin, and, therefore, the storage should not be bigger than necessary. The storage is the volume of the basin between the water levels at which the pump switches ON and OFF.

The water level at which the pump switches OFF is determined by the minimum depth to prevent air suction and excessive surface vortices. The minimum depth differs per pump. A rule of thumb is using the top of the pump housing (wet pump set-up) or the top of the suction pipeline (dry pump set-up).

It's usually not necessary to increase the shut off level in order to achieve the required Net Positive Suction Head (NPSH), as the NPSH of the pump is usually much lower than the required submergence to prevent air intake.

The water level at which the pump switches ON is determined by the lowest supply pipeline in the basin.

The minimum required storage (storage capacity), V_{req}, of the receiving basin is determined by the supply flow rate, pump flow rate, and the maximum number of pump starts per hour (source: Sewage Guideline, Module C6000 Pumping Station Management).

$$V_{req} > \frac{1}{n} \times \frac{Q_t(Q_p - Q_t)}{Q_p} \tag{4.2}$$

In which:

V_{req} = Minimum storage	[m^3]
n = Number of pump starts per hour	[h^{-1}]
Q_t = Supply flow rate	[m^3/h]
Q_p = Pump flow rate	[m^3/h]

The maximum number of pump starts occurs when $Q_p = 2*Q_t$. The necessary basin volume is:

$$V_{req} > \frac{Q_p}{4n} \tag{4.3}$$

It is obvious that the available storage must be bigger than the minimum necessary storage (equation (4.2)).

The maximum allowed number of pump starts per hour mostly depends on the switching equipment. Modern variable-frequency drives and soft starters that replace magnetic switches can deal with much higher switching frequencies. Nowadays, switching frequencies of 10 to 20 per hour are technically not a problem. Using variable-frequency drives also allows reduced pump flow rates during dry weather and in that way limit the necessary storage. However, a high switching frequency leads to energy losses because the pressure pipeline must continuously be restarted. In pumping stations with short pressure pipelines, which only pump up to the next sewage system or sewage water treatment plant, this is not relevant; however, it does apply to pumping stations with longer pipelines. The designer will have to find the proper balance.

In order to save room in the receiving basin and to prevent a splashing stream in the basin, a part of the switching storage of the pumping station can be designed in the supplying sewage pipe (inclined sump pipe).

With such a pipe, a pipeline with a larger diameter is placed deeper and under a slight slope of at least 1% after the last basin of the supplying sewage system. See Figure 4.6.

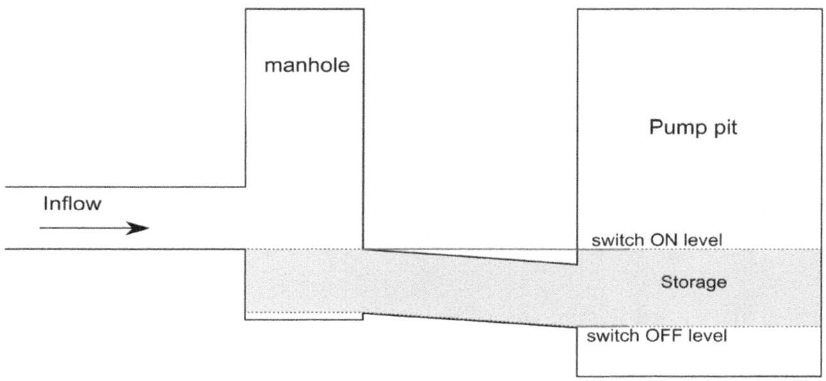

Figure 4.6 A pumping station with part of the switching storage in the 'inclined sump connection pipe' (Source: Regge & Dinkel district water board, pumping station at Enter).

This inclined sump connection pipe is used as extra switching storage in addition to the storage in the receiving basin. This way the basin can be made smaller, which is favourable for the sediment transport (less accumulation in dead corners of the basin).

With an inclined sump connection pipe, the water level at which the pump switches ON is chosen directly at the invert level of the last sewage pipe. The water level at which the pump switches OFF is chosen directly at the invert level of the inclined sump connection pipe, which avoids a splashing stream in the basin. There is enough time to discharge at the top any air bubbles that are created in the transition from the sewage pipe to the inclined sump connection pipe and thus prevent air entry into the pump.

Not all pumping stations have a receiving basin. In some cases, it might be favourable to connect the pump installation directly to the supplying pipeline, for example, in booster pumping stations. A pump installation can also be connected directly to the sewage system. However, a variable speed pump and a somewhat constant supply is necessary for this. The advantage of this method is that there is no basin where air entry and sediment accumulation can occur. The supplying sewage pipe, which must be laid deeper over the last stretch to the pump installation, is used as switching storage. This concept has been implemented in Amsterdam (Waternet).

4.5.2 Preventing air entry

In general, the water level at which the pump switches ON (and, therefore, also the highest water level) is located under the supplying pipeline in the basin. This means that the water always falls over a certain distance from the supplying pipeline. This falling water creates small air bubbles, which are sucked into the pump when the pump is operating (see Figure 4.7).

Figure 4.7 Air entry in a pumping basin (Kranendonk, 2007).

Research shows that the amount of gas that arrives in the pipeline system in this way may be substantial (approximately just as much as the amount that can be discharged hydraulically). Smit (2007) as well as Kranendonk (2007) have carried out measurements on air entry, but their results strongly depend on the water quality, pump flow rates, and basin geometry.

With simple constructions, the jet from the supplying pipeline can be broken, which substantially reduces bubble plumes under water. The literature (see Kranendonk, 2007) elaborates extensively on various constructions. The most effective constructions are a vertical plate, a T-piece with flush pipe or a rubber flap. A more detailed description can be found in Annex A.4.3.

Another logical solution to prevent air entry seems to be increasing the distance between the supplying pipeline and the pump, so that resulting bubble plume cannot reach the pump. However, this solution contradicts the other purpose: keeping the size limited and preventing sedimentation in the basin.

4.5.3 Sediment discharge and approaching flow of the pump

The fixed substances that are carried along the sewage water must be discharged as much as possible by the pumps. The design of the basin and the water level at which the pump switches OFF play an important role in this, as research has shown. Czarnota (2004) showed that a small and deep basin remains cleaner than a shallow and wide basin. By keeping the basin surface as small as possible, most of the sediment settles down near the pump and is transported when the pump is operating.

A small basin surface combined with a low water level (as low as possible) at which the pump switches OFF also results in maximum discharge of floating waste because the flow speeds in the basin will be higher and, therefore, the water at the surface will be more turbulent. As a result, more floating waste is sucked into the pump. There are various ways to keep the basin surface small and nevertheless create a large storage:

1. Deep basins with a small floor surface
2. Sloping basin walls (tapered)

The aforementioned research also shows that limiting the space between the bell mouth and walls/floor does not have much influence on pump efficiency. Most pump manufacturers provide guidelines for optimal suction of the pumps. In exceptional cases (e.g. abnormal basin design due to limited building space, large number of pumps, unfavourable connection of supplying sewage), it is wise to have a laboratory carry out a physical model test. Annex A.4 provides more information about the hydraulic design of the receiving basin.

In the past, a sequence of channels was used to carry the sediment to the pump. Such a sequence of counter-oriented channels creates a high flow speed (favourable for sediment transport), especially in the last part of the pump cycle. A sequence of channels is located above the shut-off level. A sequence of channels was used in times that switching frequency was limited by the switching equipment (magnetic switches) and start-up currents, which resulted in high engine temperatures. Pump basins became so big because of the low switching frequency that accumulation was inevitable; a sequence of channels reduced accumulation. Modern electronics has solved this problem and has made a sequence of channels unnecessary.

The disadvantage of sloping walls or fillers is that more air enters the pump (see Figure 4.8). Kranendonk (2007) provides simple and efficient solutions to prevent this.

Figure 4.8 A filler increases the air entry through the pump (Kranendonk, 2007a).

4.6 DESIGN OF AIR VALVES

All air must be removed from the pipeline system at the start of operation. This can be achieved by flushing sufficiently fast and long enough or by slowly filling the pipeline system.

Pipelines in the pumping station are usually horizontal or vertical and have a smaller diameter than the pressure pipeline (and thus higher flow speeds). These characteristics make sure that gas bubbles in the pipelines can be transported quickly (see Section 3.3 for minimum flow speeds in descending pipelines).

If air entry at the pump cannot be prevented, and the pipeline route combined with the designed speeds forms a risk for air accumulation, then a good air venting arrangement should be considered. Flow-through surge vessels (see Section A.7.5 in Annex A,

Theoretical background) can play a double role in this as water hammer surge protection *and* the venting arrangement.

A manual air valve is necessary at the pump for maintenance, because the system must refill after the maintenance work is completed.

An air valve on the pump unit is not necessary during operation if the system has been properly designed. In a properly designed system, no problematic amount of air can reach the pump or pipeline. In practice, ball check valves with a floating ball are often used as an air valve to allow air into the pipeline when the pressure becomes sub-atmospheric (after pump trip). The designer may not be aware of this and could include this arrangement because it was included in a standard design.

An air valve is not just unnecessary, it is also unfavourable. Gas bubbles can cause problems if they are transported into the pipeline (see Sections 3.2.1 and 3.3).

In the past, air valves were used to create backwash conditions so that the remaining waste could be washed out of the pump. This is not necessary with modern pumps, not even variable speed pumps with the condition that the minimum allowed pump speed is high enough, as indicated by the pump manufacturer.

Chapter 5
Dynamic effects

The design of the pipeline and pumping station must be designed for the maximum pressure that can occur in the system during non-stationary situations. Here, we distinguish between emergency scenarios, and regular and incidental operational situations. In case of manipulations with pumps and valves, as well as the behaviour of check valves and air valves, higher pressure could occur than the ones during stationary operation. There could also be a situation in which the negative pressure is low enough to cause cavitation.

Nowadays, a water hammer analysis is always carried out using a computer programme. The Joukowsky rule of thumb gives an indication of the primary pressure impulse. However, practice teaches us that this is too limited for a wastewater pressure pipeline.

The effect of water hammer can temporarily raise the pressure in the pipeline system above the permissible minimum or maximum level. As a result of a pressure surge, the pressure can temporarily become so low that it creates vapour pressure. This means that the liquid at that point converts into gas (vapour). This phenomenon is called cavitation.

In order to limit the extreme pressures, it might be necessary to install water hammer surge protections in the system. Carrying out a study for this purpose is called a water hammer analysis. A better term to use might be a 'dynamic system analysis', because this type of analysis provides more information than only on the water hammer-related matters.

An essential part of this study is a numerical simulation model. This model enables us to imitate the circumstances in normal and incidental operations. The results of this study enable us to readjust the design of the transport system.

The major added value of carrying out a dynamic analysis is that indirectly we also carry out an integral test of the design. All aspects related to pumping and pressure pipeline, as well as extreme supply patterns and outflow conditions, are included in this model.

After analysing the results, it might be necessary to adjust the design or certain components. The most important background theories for this chapter are elaborated in Annex A.7, *Water hammer in Pipeline Systems*.

This chapter discusses the design aspects related to the dynamic behaviour of wastewater in the pipelines: water hammer surge protections and control. The design process for the dynamic effects is summarised in the flow chart (Figure 5.1).

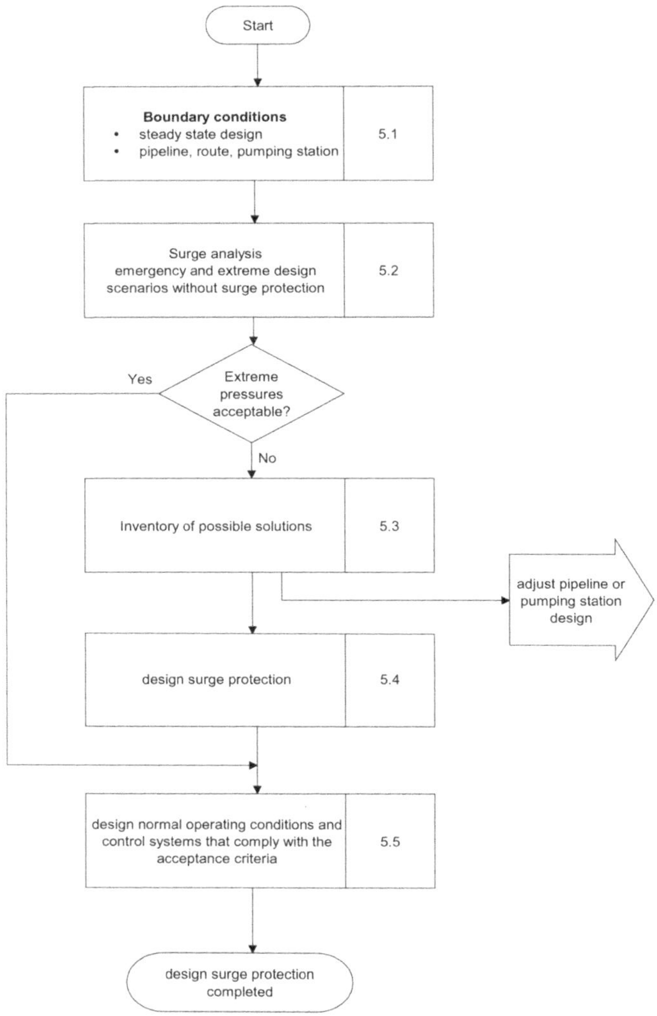

Figure 5.1 Flow chart for the design of water hammer surge protections and controls.

We stress that in case of capacity expansion or other essential changes in the system, the water hammer analysis will have to be redone. An increase in speed and possibly other pumping characteristics can considerably change the dynamic behaviour of the system.

5.1 BASIC ASSUMPTIONS

During transient operation, the most important parameters for the extent of the dynamic pressures are:

- Flow velocity changes;
- Pressure wave propagation speed (short; wave speed);

- Reflection time;
- Joukowsky pressure shock; and
- Altitude.

These parameters result directly from the pumping station and pipeline design. The maximum flow rate and the diameter of the pipeline determine the maximum speed. The maximum change in speed is determined by transient operation and characteristic of the pump or valve. In case of pump failure, the decrease in pump speed is important. This is determined by the polar mass moment of inertia of the pump and engine. When closing a valve, the closing time and the resistance characteristic that depend on the valve position (valve characteristic) are of essence.

The length and the material of the pressure pipeline are the criteria for the wave speed, the reflection time and the Joukowsky pressure change.

A pressure wave partly reflects at a change in the area and a change in the wave speed. Therefore, it is essential to know the transitions in diameters, connections to other pressure pipelines (connecting pipelines) and the use of various pipeline materials. The properties of the liquids also determine the magnitude of the wave speed. A major role in this is played mainly by the quantity of free gas in the liquid. Free gas has a reducing effect on the pressure wave speed and, therefore, on the occurring dynamic pressures. Since the quantity of free gas is not a constant parameter, the water hammer analysis uses conservative basic assumptions (gas-free liquids). In addition, the altitude of the pipeline plays a role. The highest and lowest points in the pipeline route are the usual criteria for the maximum and minimum pressures in the system.

5.1.1 Allowable maximum and minimum pressures

The allowable maximum and minimum pressures are according to selected pressure rating, other load factors (ground pressure, etc.) and the manufacturer's specifications. With thin-walled plastic pipes, there is a danger of folds (implosion) in case of high negative pressure. Make sure that the pipe connections and supports have no deviating pressure categories.

According to standards for pipelines, in general, the maximum pressure permitted for normal operating circumstances is equal to the pressure rating of the pipeline. During emergencies, the incidental pressure may temporarily be greater than the nominal pressure. Table 5.1 gives an overview of the maximum allowable incidental pressures according to various international standards.

Table 5.1 Overview of the Maximum Allowable Incidental Pressures (MAIP) in international standards.

Code	Maximum Incidental Pressure Factor [–]
DVGW W303:1994 (German guideline)	1.00
ASME B31.4 (1992), IS 328, BS 8010, ISO CD 16708:2000	1.10
NEN 3650-1:2003	1.15
BS 806	1.20
ANSI/AWWA C950-07, AWWA M45	1.40
Italian ministerial publication	1.25–1.50

Minimum allowable pressures are hardly discussed in existing standards. If the pipeline strength allows full vacuum, then the allowable pressure used nowadays is −1 barg, on condition that the implosion of the cavity is permitted, in which case a 'sharp' pressure impulse is created. The allowable cavity implosion is equal to half of the pressure category.

The reasoning behind this is as follows: the numerical modelling of the cavity and the calculation of the imploding pressures is sufficiently reliable when using validated software. After the implosion, the implosion pressure occurs for the duration of the reflection time. According to the theory of the strength of materials, this situation is considered similar to dynamic loading. In simplified structural dynamics, a pipeline with supports can be considered a single-mass spring system. The relation between the dynamic response (=tension in the pipe wall) of the system and the static response is called the Dynamic Load Factor (DLF). In the event of a dynamic load on a singe-mass spring system, the DLF is equal to two. That is why Deltares recommends using an allowable implosion pressure of 50% of the static pressure category. You can deviate from this recommendation if more complete dynamic modelling is carried out, including on specific support conditions (ground, non-linear supports), higher resonance frequency and Fluid-Structure-Interaction (FSI).

In the past, a safe limit for negative pressures of −0.7 barg was often used. There were various reasons for this; in those days, the pumps were not equipped with a speed control or with soft-stops. In the event of a pump stop, maximum negative pressure occurred. Operators wanted to remain well above the cavitation limit of −1 barg, because earlier calculation software did not contain a validated cavitation model. They also wanted to prevent degassing, which can occur in case of negative pressure. In order to keep the system well defined, they used this pressure limit. However, this is no longer necessary because validated cavitation models are widely available in pipeline simulation software.

Another boundary condition for studying the dynamic effects is to have a hydraulic model. The following section summarises the added value of a hydraulic model.

5.1.2 Hydraulic model

To study the dynamic aspects in a wastewater transport system, one must have a software programme specifically developed for this purpose. The entire pipeline system must be modelled in this programme. The schematization shows how all the pipelines, pumps, valves and other components are interconnected. Subsequently, all the relevant characteristics of these components must be included and then the calculation can begin. The pipeline system (including all components and properties) included in the programme application is called the hydraulic model.

This model can also be used very effectively for executed design steps. The programme calculates the flow rates and pressures based on the established starting criteria. The greatest benefit of using the hydraulic model in the early design stage is that the model develops alongside the design process. At the end of the design process, the designer obtains a calculation model and, with little effort, it can be utilized to determine the system's behaviour over time.

The usefulness or (better said) the necessity of a hydraulic model becomes even more relevant the more complex the wastewater transport system is.

If cavitation plays a role, then it is essential to have the calculation model use the 'invert level' (or soffit) of the transport pipeline as a reference height. After all, vapour pressure is always reached first at the top.

The hydraulic model can also be equipped with all the measuring and control devices, which enable the evaluation of the control behaviour of the system.

Finally, the hydraulic model is a useful instrument for evaluating the condition of the system during the operational phase. During commissioning of the system, the model is preferably calibrated using the as-built condition. During the operational phase, new information must be added to the model. Comparing the current measuring data with the validated hydraulic model gives the operator a reliable resource for carrying out problem analyses.

5.2 WATER HAMMER ANALYSIS WITHOUT SURGE PROTECTIONS

A water hammer analysis generally consists of carrying out a number of simulations of emergency scenarios, as well as regular and incidental circumstances, that could lead to extreme pressures. The following emergency scenarios are common for wastewater pressure pipelines:

1. Total failure of all pumps as a result of a power failure
2. Blockage of a pump or a failure of a single pump
3. Unintentional closure of a valve (if present)

Pump failure is the primary cause for most of the negative pressures in transport pipelines. If this negative pressure becomes so low that it causes cavitation then the implosion of this cavity could cause a pressure surge, which is greater than the stationary pressure.

The pump failure will cause the check valve to close at a certain moment. The closing behaviour is determined by the speed at which the flow rate at the location of the check valve decreases (the delay). In case of a significant delay, it is possible that the check valve has not yet closed at the time that the flow direction reverses. This results in a forced closure of the check valve in the return flow. This closing behaviour in turn is a water hammer phenomenon that goes hand-in-hand locally with positive and negative pressures greater than in the transport pipeline.

In the water hammer analysis, a scenario must be established in which the greatest possible delay occurs in the check valve. In a pumping station in which two or more pumps operate, this is a failure of one pump. For more information about the dynamic behaviour of check valves, see Appendix A.6.3.

Rapid closure of the valves can lead to impermissible positive and negative pressures. A wastewater transport system rarely has control valves that are part of the operating process. Opened/closed valves in principle can only be operated in case of a 'pump stand-by' situation. If there are situations in which valves can be manipulated in flow conditions, then the consequences of these must be analysed beforehand.

The input parameters used in the simulation must be selected specifically to reach extreme values. This is called conservative modelling. It is very common to carry out calculations using clean water properties as opposed to wasterwater properties, even though wastewater is far from clean. Clean water properties are oftern utilized because they provide the most conservative result. Free gas (in the form of gas bubbles) in wastewater reduces the wave speed and thereby the magnitude of the pressure impulse.

When evaluating the results and the need for a surge protection, it is recommended to consider potential consequences (therefore risks).

A water hammer analysis, in addition to identifying the occurring pressures, also provides insight into the dynamic functioning of the entire pipeline system. Two examples specifically mentioned are:

1. Following pump failure, supply might still continue for a considerable time because the check valve does not shut immediately. This continued supply might reduce the level in the suction basin even further and create the conditions for air intake.
2. In a pressure pipe system with connecting pipelines, a negative pressure wave initiated by one pumping station might cause siphoning in another non-operating pumping station. Draining out the suction basin is a risk here.

The simulation model is a simplification of a practical situation. There are always physical processes that are taken into account in a simple manner or not at all. The choice of conservative model parameters can differ per scenario. It is recommended to analyse the influence of uncertain model parameters (for example, wall roughness) on the calculation results.

In many cases, the emergency scenarios without surge protections lead to impermissible dynamic pressures. In these cases, a list of possible solutions must be made. The solutions could consist of system adjustments, for example larger diameters, water hammer surge protections, emergency controls or a combination of these solutions.

5.3 INVENTORY OF SOLUTIONS FOR PREVENTING IMPERMISSIBLE PRESSURES

If impermissible pressures occur, the designer can choose from four solutions in order to limit water hammer:

1. Adjust the diameter, the pipeline route or profile
2. Adjust the wave speed (pipe material, wall thickness)
3. Adjust the flow velocity change
4. Apply a water hammer surge protection that will limit the consequences

Option 1 and 2 are usually possible only if an exploratory water hammer analysis was carried out in an earlier phase and there is still manoeuvring room in the design of the pipeline trajectory. Option 3 and 4 affect the design of the pumping station.

Option 1

High points in the pipeline route generally determine the locations in which the lowest pressures occur following pump failure. If these locations show impermissible pressures and the design still allows for changes in the installation, then alternative routes can be considered. Another possibility is to install the pipeline deeper at those locations. The technical costs and the rest of the measures that this would entail will indicate whether this option is feasible.

Option 2

To reduce the wave speed and the related dynamic pressure surge, a more elastic pipeline material will have to be applied. If the designer still has the flexibility to do so, an alternative should be explored.

Option 3

Reducing the flow velocity changes reduces the also dynamic pressures. In case of pump trip, speed controller or soft starters are options. In this case, however, it should be noted that these electro-technical provisions no longer have any effect in case of power failure.

In case of pump failure, the steepness of the pressure wave is determined by the magnitude of the polar mass moment of inertia (I_p) of the engine, pump shaft and impeller. By selecting another pump/engine, the polar mass moment of inertia can possibly be increased. Applying a flywheel is not a realistic option in the case of pumps that frequently start and stop.

In case of impermissible pressure caused by manipulation of a valve, the closure time must be increased. Another possibility is to choose a non-linear closure pattern. The flow reduction in case of a constant closure time of a valve does not actually run linearly. By closing the first trajectory faster and the second trajectory slower, the maximum pressure can be limited. Another solution is to use a stepper motor, which creates somewhat of a waiting time between two valve positions. Effective closure time also plays a role for a valve. Depending on the valve characteristics, the flow rate in the beginning of the closure trajectory will hardly be affected. With butterfly valves, for example, a flow rate change is observed only in the last 30% (see Figure 5.2).

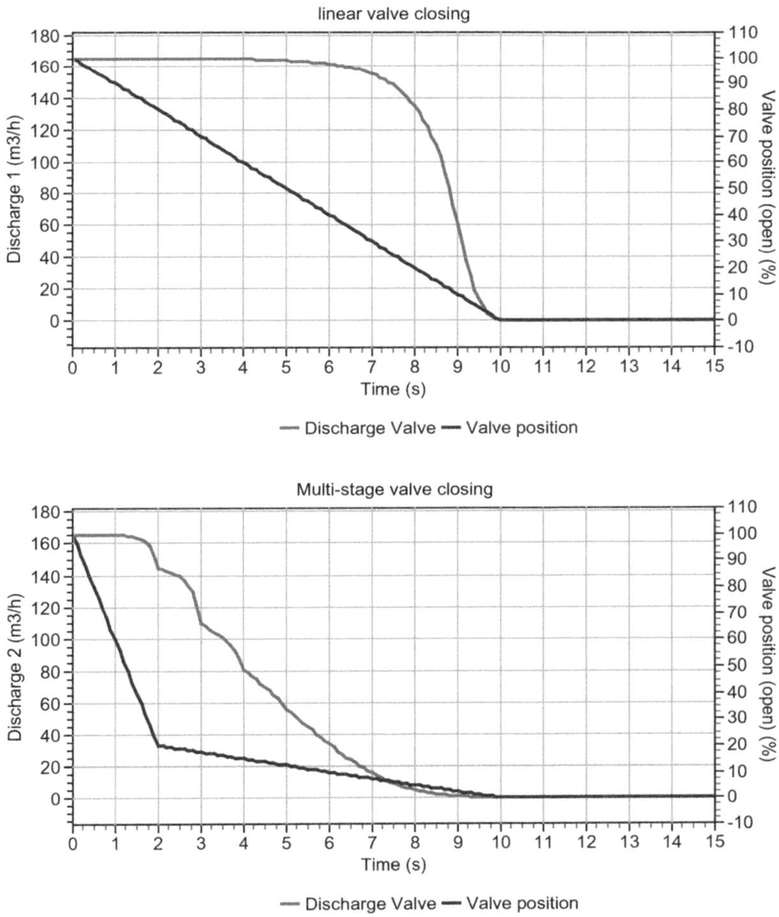

Figure 5.2 Effect of closure behaviour on flow reduction. The top figure shows a linear closing valve, the lower picture shows a valve closing in two steps.

When removing air from an empty pipeline (or part thereof), be careful that the filling speed does not get too high. For example, the pump should be started at a low pumpspeed, to prevent a high flowrate and thus high filling speed. As soon as the gas pocket is expelled, the moving column of water slams into the (almost) still-standing water column. Filling the system, therefore, requires advanced planning.

Option 4

If the aforementioned measures are impossible, then a water hammer surge protection will have to ensure that the consequences of the disturbance remain limited. Using a water hammer simulation programme, first make a list of which types of surge protection will be considered. Subsequently, the chosen type of water hammer surge protection will be further dimensioned. The principle here is that the surge protection must supply water long enough to prevent impermissible pressure.

5.4 DIMENSIONING WATER HAMMER SURGE PROTECTIONS

The following surge protections might be considered:

1. Surge vessel, usually combined with a fast-closing or dampened check valve
2. By-pass pipeline
3. Air vent
4. Surge tower

In appendix A.7, the attention points are elaborated further per surge protection. Surge towers are hardly used any more due to their height and stench problem, but since they are still in use, for the sake of completeness, they are mentioned in the appendix.

A rule of thumb can be used when dimensioning the surge vessel in order to determine the necessary surge vessel volume. For wastewater pressure pipelines in flat areas, the formula (5.1) provides an indication of the necessary surge vessel volume.

$$V = c \cdot QL \tag{5.1}$$

In which:

V = Necessary surge vessel volume	[m^3]
c = Constant, 0.01 to 0.02	[s/m]
Q = Maximum flow rate	[m^3/s]
L = Length of the wastewater pressure pipeline	[m]

A surge vessel (and surge tower) greatly affects the deceleration of the flow at the location of the check valve. In order to determine which type of check valve is necessary in connection with closure during return flow, a detailed model of the pipeline system is necessary between the suction chamber and the surge vessel or the surge tower. Reflections or closure of the check valves in the return flow can easily cause cavitation in the pipelines or the pumping station.

Nowadays, many sewage-pumping stations are constructed with variable speed drives that control the pump speed. The control ensures that in case of pump stop, the revolutions drop slowly to zero. Another electro-technical method is the use of a soft starter instead of a star-delta switch. A soft starter is mainly intended to level the starting current. The soft starter can also be used as a 'soft stopper'. This allows you to adjust the revolution increase and decrease over time. Both electro-technical provisions are in fact water hammer surge protections because the speed variations and the related pressure fluctuations are much smaller in comparison with ON/OFF pumps in which the engine is switched off immediately.

5.5 NORMAL AND INCIDENTAL OPERATIONS

The following scenarios are considered normal operations (also see Appendix C.2.2. of NEN 3650-1:2003):

1. Pump start with a filled system
2. Controlled pump stop
3. Acceptance tests; these may sometimes be more extreme than emergency scenarios.
4. Filling procedure or a pump start in a partly filled system
5. Procedure for emptying a pipeline system (or part thereof) for the purpose of carrying out maintenance work
6. Controlled valve closures
7. System with several pumping stations: Stop of (fixed speed) pumping station A and start of pumping station B
8. Other manipulations that bring about delay or acceleration of the flow
9. Switching procedures
10. Effects of faster controls (relevant only for systems that do not have open storage, such as booster systems)

Normal procedures must not lead to the activation of emergency controls; otherwise, the emergency controls have to be adjusted. Finally, it might be necessary to devote attention to special circumstances, such as:

1. Deviating use of the pressure pipeline
2. Pump start with air or gas pockets; this may cause extreme peak pressures.
3. Faulty switching procedures

Simulating the normal operating situation provides insight into the behaviour of the system. Depending on the complexity of the system, it might be necessary to carry out several simulations. Three common situations are:

- System operations under dry weather conditions,
- System operations under wet weather conditions,
- System without pumps in operation (mainly for checking static pressure in the pipeline)

In dry weather, the ON/OFF operation of the pump(s) is primarily checked. An essential factor here is that the hydraulic model supports a realistic pump control.

The aspects that have to be analysed are:

- How fast does the flow rate increase?
- What are the minimum and maximum occurring pressures?
- Switching cycles of the pumps

- Do the set control parameters ensure the correct behaviour?
- Levels in the chambers, basins, etc. (important during temporarily continued supply after pump stop)

In wet weather, conditions are simulated in which the supply increases over time from zero-supply to wet weather supply with a speed that the designer considers to be realistic.

The aspects that have to be analysed are:

- Can pump control maintain the supply increase well enough?
- What are the minimum and maximum occurring pressures?
- Maximum levels in chambers, basins, etc.

If the wastewater transport system uses automatic control valves, then their closure is also a situation that can be analysed. This is definitely necessary if the valve takes over the function of a check valve.

Carrying out regular operating management situations provides a timely insight into the dynamic behaviour of the system. This is useful information at the start of operations. For example, by comparing the simulated time series of the flow rate and pressure in the pumping station with the real situation, one can evaluate whether the system is properly vented.

In pressure pipeline systems with connecting pumping stations, one must examine the other pumping station with regard to mutual inter-action. As a result of reflecting pressure surges, other non-operating pumping stations can subsequently deliver enough flow, causing their receiving basins to drain and air being sucked into the system.

In pumping stations that are not equipped with a frequency converter or soft starter, significant dynamic pressures also occur during regular operations. With regard to such pumping stations, the contents of Section 5.2 are even more significant because the water hammer phenomena can occur during any pump switch off.

A designer of complex pressure systems is often forced to play *the devil's advocate* in order to define the scenarios that might provide a reasonable chance of prevention and lead to extreme pressures decisive for the design.

Chapter 6
Design aspects for maintaining capacity

The design phase must devote attention to parameters that are necessary to evaluate whether the system achieves the design capacity. On the one hand, these parameters are necessary during delivery and acceptance of the system. On the other hand, they are necessary during the utilisation phase for periodically determining the capacity of the system. These parameters must be defined so that after reducing the capacity, the level can be brought up to the design level.

To characterise the capacity of the system, a performance indicator (PI) is introduced. Measurements are necessary in order to determine and maintain the PI. The parameters that can help to decrease additional energy losses in the pressure pipeline are mentioned last.

6.1 PERFORMANCE INDICATORS

A wastewater pressure pipeline system is designed to transport a certain flow using a certain amount of energy. In the course of time, the system could deviate in the way it operates. In an ideal situation, the operator notices these deviations well in advance and takes necessary measures to address these deviations. In practice, however, things are often different. Proper functioning is often measured by the capacity or, even more rudimentarily, by "does the sewer system overflow?".

Pumping stations equipped with flow control can always achieve this flow rate as long as they do not exceed maximum pump speed. This goes hand-in-hand with generating additional capacities; in other words, simply looking at the flow rate or the water level in the basin is not sufficient. That is why the general term 'performance norm' or 'performance indicator' (PI) was introduced. In order to assess whether the PI is being achieved, it first has to be known what the system is meant to do.

The performance indicator (PI) is the ratio between two parameters, $K_{reference}$ and $K_{real\ time}$. The parameter $K_{real\ time}$ is based on the measured data gathered over time. Each parameter has a reference value, determined in the design or during the start of operations, and a real time value. The performance indicator is the ratio between the real time value and the reference value. The ratio is calculated in such a way that reduced functioning leads to a PI < 100%. A PI should become an average over the course of time. This is the only way to compare different systems with the same parameter ratio (PI). The selected PIs can always become an average (weighted) over the various systems so that a PI can be achieved per region/municipality/water board/operator.[1] Each system can have several parameters in order to define the objectives. Examples of parameters are:

- Operating hours per period (day, week, month, year)
- Operating hours per m^3 per unit time
- Required pump speed for the flow design (for pumping stations regulated according to revolutions)
- Specific energy consumption (kWh/m^3, averaged over a day/week/month/year)
- Transport capacity
- Pump efficiency
- Resistance factor of the transport pipeline or system characteristic
- Equivalent wall roughness
- Pump characteristic (i.e., operating point with regard to noticing wear and tear of impeller)

In new systems, it is very easy to determine the reference value $K_{reference}$ during the acceptance test. Usually, this will comply with the design conditions. In existing systems, a new baseline-measurement ('benchmark') has to be carried out. It is then essential that the state of the system is established unequivocally.

In addition, the design basis must be set to this zero-measurement so that in new projects the designer can assume a realistic design basis. Effectively, this means that the designer has to understand the system characteristics under the recognised conditions of a normally functioning system. Together with the data of the installed pump and with help of records, we can verify whether over time the system will start deviating from its starting reference values. This conditions monitoring can be effectively incorporated into a monitoring system in which the manager is informed only if the deviation becomes too large. However, at the same time, periodic manual processes and evaluations can also be used.

The $K_{reference}$ is a feature of a specific pipeline system and it cannot simply be compared to other pipeline systems. However, the PI has no dimensional value (i.e. it is a percentage) of the actual parameter value and reference value. As soon as this deviates from 100%, it means that something in the system has changed. This makes it possible to compare the PI with the PI of another system.

The cause of a capacity reduction might be attributed to the pump installation or the pressure pipeline or both. The parameters described below identify the entire system but do not indicate immediately what the cause of the capacity reduction is. For this purpose, we have to rely on additional measurement data (depending on the method), such as pump speed or outlet pressure.

The user must be aware of the limitations of the parameter used. Not all parameters can be used for all objectives. Below are several elaborations of parameters and PIs that can be applied. A more elaborate document has been made in collaboration with Stowa for the practical implementations of performance indicators in SCADA systems (Deltares, 2013).

6.1.1 Operating hours

For smaller pumping stations (without variable speed pumps), the operating hours of the pump per unit time (e.g., month) can be easily used to determine the parameter.

---

[1] If there is an agreement within a district water board about the reference values, the department/region managers can be rewarded for improving a PI in order to stimulate involvement in the functioning of the transport systems.

The advantage of this parameter is that the operating hours of the pump are generally updated for the purpose of maintenance. Therefore, additional instruments are not necessary. In practice, additional instruments are often not available. During the design phase, based on the expected supply and the installed pumping capacity, we can estimate the number of operating hours. This type of parameter is appropriate for simple pressure sewage systems but not for pumping stations that are connected to a mixed sewage system. In that case, in order to interpret the parameter, we would have to know the precipitation amount. The disadvantage of operating hours is that the operating hours of the pump do not provide any indication of the cause of the problem. The increase in the number of operating hours might also be on the supply side. Therefore, this indicator can only be used for determining that there is a problem, not for determining the problem.

6.1.2 Specific energy consumption

Specific energy consumption is understood to mean the quantity of energy required to transport one m^3 of water. The same parameter can be determined in various ways.

$$K = \frac{E}{V} \frac{[kWh]}{[m^3]} \tag{6.1}$$

By dividing both terms by one hour, you get an identical parameter, but now expressed in power P and flow rate Q:

$$K = \frac{P}{Q} \frac{[kW]}{[m^3/h]} \tag{6.2}$$

The elaboration of the capacities shows that primarily the pump head determines this parameter:

$$P_{elec} = \frac{P_{hydr}}{\eta_{pump} \times \eta_{elec}} = \frac{\rho g Q \Delta H}{\eta_{pump} \times \eta_{elec}} \tag{6.3}$$

In which:

P_{elec} = Electrical power	[W]
P_{hydr} = Hydraulic power	[W]
η_{pump} = Pump efficiency	[-]
η_{motor} = Engine efficiency	[-]
ρ = Density	[kg/m^3]
g = Gravitational acceleration	[m/s^2]
Q = Flow rate	[m^3/s]
ΔH = Pump head	[m]

The electrical power can be calculated directly from the power consumption and the voltage. Often, only the power consumption of the pump is registered because the voltage is considered constant.

$$P_{elec} = U \cdot I \tag{6.4}$$

In which:

U = Voltage [V]
I = Current [A]

The energy consumption of the sewage pumping station is updated automatically by the energy supplier, which makes the application of equation (6.4) easier in practice.

However, this registration contains the total energy consumption of the sewage pumping station, therefore, including lighting and cooling. For the correct use of the equation (6.4), a correction would have to be applied.

If we relate the parameter only to the hydraulic power, then only the pump head emerges as the essential factor for the specific energy consumption:

$$K = \frac{P_{hydr}}{Q} = \frac{\rho g Q \Delta H}{Q} = \rho g \Delta H \tag{6.5}$$

$$PI = \frac{K_{reference}}{K_{actual}} = \frac{\left(P/Q\right)_{reference}}{\left(P/Q\right)_{actual}} = \frac{\Delta H_{reference}}{\Delta H_{actual}} \tag{6.6}$$

In complex pressure pipeline systems, the delivery pressure can vary considerably depending on whether it operates with or without other pumping stations. To be able to determine the correct PI value, the reference value of the current condition must be known. The current condition of the system must, therefore, be registered. The graph below shows an example of the current Q/P ratio for various flows and pump operations at three moments in time (week 0, 4 and 8).

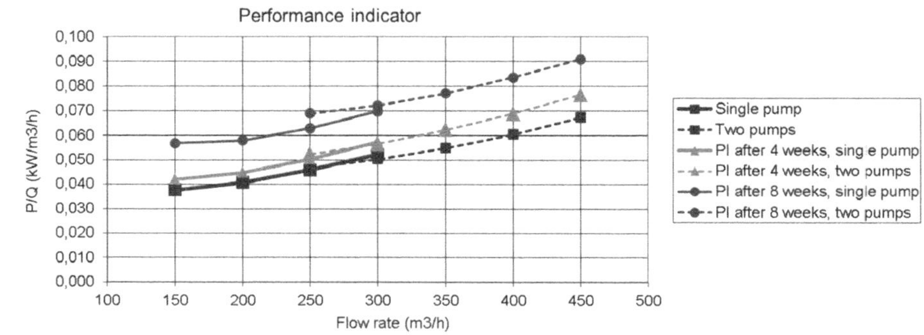

Figure 6.1 Example of a parameter, based on flow rate and capacity.

In case of a condition that occurs frequently in the system, the PI can also be set up over time in order to quickly evaluate whether performance dropped to a critical value in which it is desired or necessary to intervene.

This parameter can be evaluated on various time scales.

Short time scale (in the order of minutes)

This parameter is not constant, but dependent on the flow rate, static pump head and the number of operating pumps. Therefore, when determining the correct PI value, the correct reference value belonging to the present condition must be used in order to determine the PI at that point in time.

Long time scale (in the order of weeks/months)

For developing the specific capacities for the long term, we have to use a reference value from the historical data. Such a fixed reference value allows us to see whether the average pump head changes over the course of time. Adjusting the control (for example, not operating the pumping stations simultaneously whenever possible) may result in a lower average pump head, which will increase the PI.

6.1.3 System and pump characteristics

A PI for a single pressure pipeline can also be expressed as the relationship of pipeline losses. Pipeline losses in the reference situation ΔH_{ref} can be modelled using equation (6.7), in which the energy loss is calculated by the difference between the head of the discharge pressure in the sewage pumping station H_{pg} and the fixed or measured suction pressure head H_p.

$$\Delta H_{ref} = H_{pg} - H_p = C_1 Q^2 + C_2 Q \tag{6.7}$$

At each measured flow rate and discharge pressure of the sewage pumping station, the PI is calculated as follows:

$$PI_{\Delta H_{leiding}} = \frac{\Delta H_{ref}(Q, H_p)}{H_{pg,act} - H_p} \tag{6.8}$$

Measuring the discharge pressure and suction level can ultimately be used as well for determining the pump head and for comparing the pump curve. The actual pump head is derived via equation (6.9)

$$\Delta H_{pomp,act} = H_{pg} - H_z + C_g Q^2 \tag{6.9}$$

in which, C_g represents the local losses in the sewage pumping station. This allows us to calculate the PI for the sewage pumping station as follows:

$$PI_{\Delta H_{pomp}} = \frac{\Delta H_{pomp,act}}{\Delta H_{pomp,ref}(n, Q)} \tag{6.10}$$

These two performance indicators can be shown graphically in a Q-H diagram. In the example in Figure 6.2, the theoretical operating point will be at 20 m pump head and 440 m^3/h flow rate. The difference between the actual measured operating point (red marker in Figure 6.2) and the system characteristics means that there is additional resistance in the pipeline. The difference between the pumping curve and the measured operating point indicates that the pump is operating less efficiently (for example, due to damage to the impeller or a mediocre approaching flow).

The advantage of the latter two performance indicators is the direct relationship to the energy efficiency, because the energy consumption runs proportionally to the pump head.

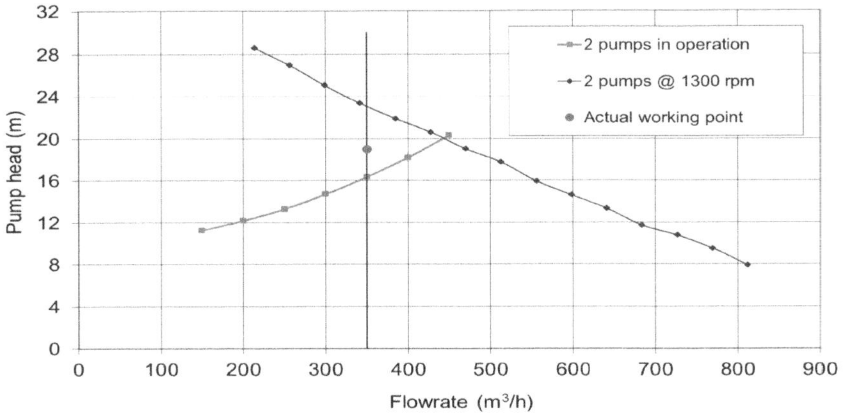

Figure 6.2 Operating point, pumping curve and system characteristics in Q-H diagram.

6.1.4 Equivalent wall roughness

The equivalent wall roughness appears to be a simple and easy parameter to use. However, it is a calculation quantity that depends on the friction factor (λ), diameter and the Reynolds number (see Annex A.4).

The problem with wall roughness is that large pipelines have an advantage: Figure 6.3 shows the relationship between the Darcy-Weisbach friction factor and the pipeline diameter (during flow speed of 0.5 m/s).

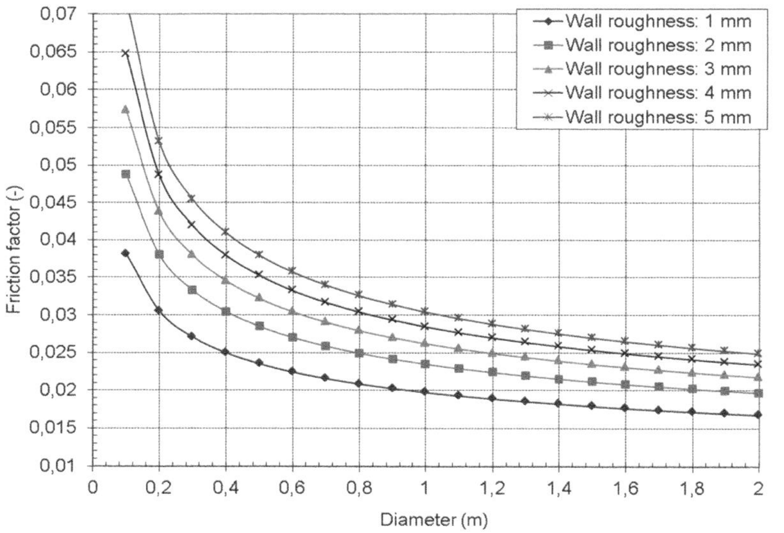

Figure 6.3 Friction factor of the pipeline diameter for several types of wall roughness.

Example: A wall roughness of 5 mm in a Ø1500 mm pipeline gives the same resistance coefficient λ as a wall roughness of 1 mm in a Ø300 mm pipeline.

Knowing only the wall roughness gives no indication of the pipeline condition if the diameter is unknown. In practice, there may be minor variations in the diameter due to growths, scaling and slime layers on the pipeline wall.

Therefore, this parameter cannot be transferred to other pipeline diameters. However, wall roughness can be compared to other pipelines with the same diameter, but with different lengths. The wall roughness is also not sensitive to static pump head.

The major difference with the parameter as described in Section 6.1.2 is that the 'k-value parameter' only says something about the condition of the pressure pipeline and nothing about the total system, including the pumping station.

Example of PI determination based on the k-value:

Reference k-value = 0.25 mm

Actual k-value = 0.31 mm

$$PI = \frac{0{,}25}{0{,}31} * 100\% = 81\%$$

6.2 MEASURING LOCATIONS FOR PRESSURE AND FLOW RATE

The cause of capacity reduction can be traced to the pumping station and/or the pressure pipeline. With suitable measurements, one can determine the origin of the problem. It is, therefore, important to include all the relevant measuring points in the design process. The energy losses can be identified by determining the pressure differences and the flow rates. In order to figure out the cause of the capacity reduction or increased energy consumption, we have to distinguish between the performance indicators for the pumping station and for the pressure pipeline. A pressure gauge on the delivery side of the pumping station is essential in order to observe the differences. Traditionally, such an outlet pressure gauge is not installed, but it can provide a wealth of additional information about the hydraulic functioning of the transport system.

In this manual, we have decided to leave aside the required measuring values for the operational management. In practice, several measuring values (for example, flow rate and suction level) are used for both applications.

Many manufacturers of measuring instruments set down requirements for installing these instruments. Not following these requirements can seriously affect the accuracy of instruments. For practical reasons, people may find that they must deviate from physically ideal measuring locations. A hydraulic model might also help here in making a sensible 'second choice'. In newly-built situations, it is recommended to take into account the option of measuring, both with respect to normal operation and – for example – for the purpose of periodic tests, for the presence of gas pockets.

6.2.1 Flow rate measurement

A precise measurement of the flow rate can only be obtained by using a flow meter. An electromagnetic flow meter (EMF) is normally applied because this type does not have additional resistance or any moving components (and is, therefore, reliable).

The manufacturer specifies the accuracy of the flow meter. The flow meters are frequently calibrated using clean water, which may result in a reduced accuracy in wastewater.

Accuracy is also affected by the installation location. There should be sufficient uniform flow available upstream and downstream of the flow meter. These required dimensions are fixed in the norms and are specified by the manufacturer. International standards such as ISO-5167:2003 provide values of ∅10 to uniform length upstream of the flow meter and ∅2 to uniform length downstream of the flow meter in order to rule out the effects of non-uniform flow. In practical situations, this is not feasible. In consultation with the manufacturer, other lengths can be used.

If flow meters are not installed (which is the case in old pumping stations) the flow rate must be determined by pumping out a defined amount of water from the suction basin during a certain period. In this case, the supply to the chamber must be closed or information about the filling time of the chamber must be used to estimate the supply. These measurements are not accurate because often it is not possible to continue pumping long enough to obtain a steady flow in the pipeline. During such operations, surge vessels and buffer towers must be closed because otherwise one measures the partial filling of the vessel/tower and not the capacity of the pipeline.

6.2.2 Pressure measurement

To determine the energy loss of an object, one must register the upstream and downstream pressures of this object (see Figure 6.4). Therefore, at least between the pump and the check valve, or at the beginning of the pressure pipeline, it is worthwhile to install a connection point for a pressure sensor or manometer on the pipeline (S3, S4 or S5 in Figure 6.4).

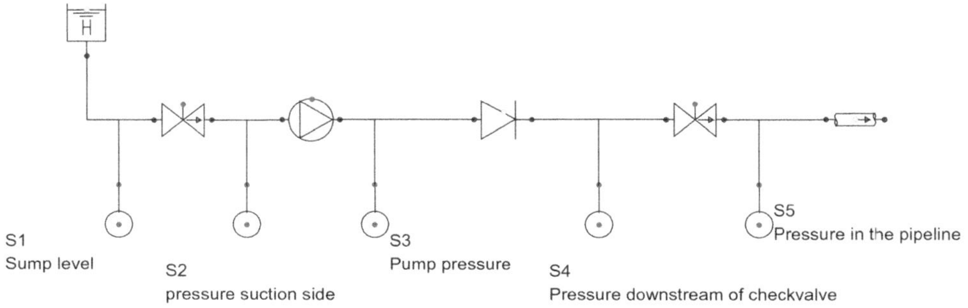

Figure 6.4 Locations for measuring pressure.

In addition, the suction level in the pump sump (S1 in Figure 6.4) must also be registered. This pressure registration is usually used for switching the pumps on and off. The necessity to register the delivery pressure level (pressure at the end of the system) depends on the discharge construction. This is often a weir-like construction in which the level at the end of the system is known.

When installing dry pumps, a pressure-measuring point can also be installed on the suction side (S2). This allows us to evaluate immediately whether a blocked suction mouth or an incorrectly functioning pump causes the capacity problem.

The best location for pressure gauges is on the axis (side) of the pipeline. At this location, there is less chance of blockage by sediment or floating waste. International standards such as ISO-5167:2003 also provide correct pressure measurements necessary for uniform lengths: ∅2 before and ∅2 after the measuring point.

As a result of limited pipeline length in the pumping station, there is often no compliance with this in practice. It is recommended to interpret the measurements carefully. Especially locations immediately after the check valves cause a too low pressure-registration because the measuring takes place in the vena contracta of the valve.

The height level of each measured pressure point must be accurately determined and fixed. This level is necessary for the correct analysis of the measurement.

When determining the energy loss based on measuring pressure differences, with regard to deviating diameters, one must take into account the difference in velocity head (see appendix A.3).

To determine the performance of the pressure pipeline, it is minimally necessary to know the pressure at the beginning and at the end of the pressure pipeline. If certain pressure pipelines are susceptible to extra energy loss – due to the geometry, operation and wastewater quality – then it is desirable already during the design phase to include additional connection points along the route. These measuring points can be used incidentally in the operating phase if additional information is required. Additionally, it is recommended to include at the air valve location a direct connection point for pressure measurement.

Connecting pipelines that are not in operation can be used to determine the pressure at the location of the connection point. For this purpose, a measurement point must be installed at the beginning of the pressure pipeline (therefore, behind the check valve).

6.2.3 Pump data

In order to evaluate the performance of variable speed pumps, the following parameters must be known: flow rate, pump head, pump speed and consumed power.

6.3 PROVISIONS FOR MAINTAINING CAPACITY

The cause of capacity reduction may be attributed to fouling / blockage of the components in the pumping station or fouling and/or stagnating gas bubbles in the pressure pipeline.

6.3.1 Valves

To solve the capacity problem in the pumping station, it must be possible to inspect/dismantle the pump and/or the check valve. For this purpose, the pipeline section needs to be shut off from the suction basin and pressure pipeline using valves. To ensure an undisturbed cross-sectional area for the flow, it is recommended to use gates or gate valves.

6.3.2 Increased flow speed

To remove a gas bubble by means of flow, the flow must have sufficient speed. The required flow speed depends on the slope angle and the diameter of the declining sections of the pipeline (see Section 3.3 and Appendix A.8). There must be sufficient pumping capacity to realise this speed.

Removing gas bubbles easily takes several hours. The figures in Appendix A.8 can be used to determine the required degrading time. During this time, sufficient water reserve must be guaranteed. This can be achieved by:

- Sufficient buffer in the receiving basin
- Using the upstream sewer system as a buffer
- Supply from nearby surface water

If this cannot be achieved, the removal by means of a flow is impossible and the options are then limited to pigging and venting.

6.3.3 Pigging

A pigging facility must be installed at the beginning of the pressure pipeline. The most simple method is a Y-piece with a disc valve and blind flange (see Figure 6.5). The Y-connection at the beginning of the pressure pipeline can also be used for connecting a temporary pumping installation in case the pumping station cannot be operated (due to renovations or an emergency).

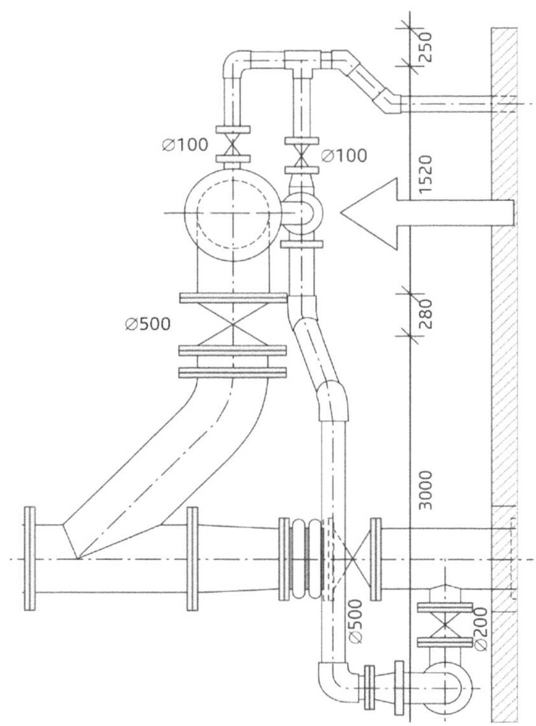

Figure 6.5 Top view of a launching station for pigging.

A launching station must be equipped with a connection for increasing the pressure behind the pig (by using a pump or a temporary installation).It is recommended to install a departure signal. This can also be done by registering the flow rate and pressure.

The travelling time of a pig depends on the flow speed (corrected for the extra resistance of the pig). During this time, there has to be sufficient water reserve.

Pigging a pipeline imposes certain requirements with regard to the change in diameters inside the pipeline. This depends on the type of pig used. In turn, the type of pig used depends on the nature of the waste. For removing gas and cleaning of non-industrial pipelines, it is usually sufficient to use a soft over-sized pig (the diameter of the pig is somewhat bigger than the diameter of the pipeline). This type of pig can be deformed so that it can safely push through smaller diameters up to $0.7D$ and bends with a radius larger than $1.5D$ (in which the diameter, D is the smallest pipeline diameter in the pipeline system).

Chemical and biological processes in industrial wastewater can cause hard deposits on the wall (scaling), which can be removed only with a hard pig. With this type of pig, there cannot be any deviation of pipe diameter throughout the path.

If the pipeline system has obstructive elements, such as butterfly valves, then pigging is not an option unless the system is adapted. If there is free flow to a receiving basin, then a separate receiving installation is not necessary. There is, however, the need to have the pig hoisted out of the receiving basin.

6.3.4 Air valve locations

With regard to locations in the pressure pipeline where stagnation of gas bubbles could occur, reference is made to Section 3.2.1. If the height difference is great enough that significant energy loss might occur (maximum energy loss is energy loss the pump can overcome), a venting provision must be installed. To achieve good venting, an upright pipe section must be mounted in compliance with Section 3.4.2.

For air valves in the pumping station, refer to Section 4.6. For incidental air release (for example, after maintenance work) of the pump and/or pressure pipeline, we advise to opt for manual air valves.

Chapter 7

Review of the overall system design

In the introduction of chapter 2, it had been stressed that the design process is an iterative process. The design process cannot be divided into a number of sub-designs that are implemented independently. One must constantly examine the consequences of certain decisions on the rest of the system.

In the preceding chapters, these interrelationships were established in which the core issue is the transport of gas. Section 7.1 summarises a number of design and operational measures for controlling or preventing air in the pressure pipelines.

For the purpose of optimal coordination between the various design steps, at the end of the design track it is wise to check and test the entire system design with several people from various disciplines to verify whether the final design complies with the various aspects.

Particularly when making adjustments to existing systems, one might overlook the effect of the adjustment on the functioning of the rest of the system. The 'review of the integral system design' is actually just a checklist with a number of attention points per object. The tables in Section 7.2 provide the first attempt to review the system design.

7.1 CONTROLLING THE AIR IN PRESSURE PIPELINES

Table 7.1 Overview of potential measures for controlling the problem of air.

Phase, Sub-system	Measure	Comments
Design, Pumping station	Install a vertical splash shield, a vented T-piece or a rubber flab in order to guide the flow of wastewater in the basin.	This measure is highly recommended and is very effective if the waterfall entrains into the basin near the suction mouth.
	Install a pigging installation consisting of a Y-piece or a by-pass with isolation valves.	A pigging action can also be used when initially filling the pipeline.

(*Continued*)

Table 7.1 Overview of potential measures for controlling the problem of air (*continued*).

Phase, Sub-system	Measure	Comments
	Close venting (air release) pipelines in the pumping station.	In many pumping stations, these venting pipelines are made with small ball-check valves that unintentionally allow air to enter during negative pressures.
	Install instruments for monitoring the energy efficiency of the pumping station and pressure pipeline.	Flow rate, suction level, outlet pressure, pumping revolutions and consumed capacities/power are recommended for calculating real-time performance indicators for the pumping station and pressure pipeline.
Design, Pipeline	Limit the number of crossings.	A directional drilling is maximum around 1 km long, in which sometimes several crossing pipes can be combined.
	Limit the depth of a crossing pipe.	The feasibility depends on the geo-technical situation and the structure of the substrate.
	Reduce the diameter of the crossing pipe to increase the local flow rate (and the flow number).	Permanent extra energy loss: Pigging with flexible pigs should still be possible; the result is that the diameter can sometimes be reduced by 30% to 50%.
	Reduce only the interior diameter of the sloping leg of a crossing pipe.	This is possible by opting for a smaller wall-thickness-diameter-ratio or by carrying out the drilling against the flow direction, so that the larger diameter is first pulled through the drilled hole.
	Install manual or automatic air vents located upstream of the sag pipes.	Maintenance is required.
	Replace a directional drilling with a traditional crossing pipe.	
Operational measures, Pumping station	Raise the water level at which the pump switches OFF.	Based on a water hammer calculation, the required level is calculated, so that a decrease in level does not lead to intake of air after a pump stop.
	Improve the stop procedure with a soft-stop.	This limits the minimum pressures during normal operation. A water hammer calculation is necessary to design an acceptable stop procedure.
	Install a return pipeline in the pumping station from the discharge side to the basin. This pipe is initially open and closed shortly after the pump start, so that any air can be pushed back to the basin.	The pipe section between the pump and the check valve normally stands at negative pressure during stoppage, so that the air and gases can escape out of the water.

7.2 CHECK LISTS

7.2.1 Pumping station design

Table 7.2 Design aspects and potential measures for the pumping station.

Pumping Station Design Aspect	Potential Measures
Is there a chance of air entering the pump as a result of falling influent water?	**Yes:** Place a curtain wall. **No:** OK
Is the shut-off level sufficiently high so that sufficient water can be delivered after pump stop without the risk of air suction?	**Yes:** OK **No:** Raise the level.
Does the water level at which the pump switches OFF come out under the invert level of the sewer so that incoming water can break up the floating layer?	**Yes:** OK **No:** Lower the level.
Is it possible to clean the basin with a suction vacuum truck, for example?	**Yes:** OK **No:** Adapt the design.
Is the pump switch storage big enough?	**Yes:** OK **No:** Adapt the levels or make the basin larger.
Are air valves on pump closed during normal operations?	**Yes:** OK **No:** The risk is undesired air intake during negative pressure. Close the air valve.
Is the pipeline resistance in the pumping station included correctly in the design? (The position of the check valve need not be entirely open.)	**Yes:** OK **No:** Include extra dynamic loss in the design.

7.2.2 Design of pressure pipelines

Table 7.3 Design aspects and potential measures for pressure pipelines.

Pressure Pipeline Design Aspect	Potential Measures
Is the pipeline system entirely pressurized if all the pumping stations in the system are switched off?	**Yes:** OK **No:** Make a list of possible consequences of air pockets. Take measures if the risks are unacceptable.
Are there locations of gas pockets along the route that might cause large additional head losses?	**Yes:** Test the controllability, if necessary, take measures. **No:** OK
Are there provisions made to remove air at certain locations?	**Yes:** OK **No:** Test the controllability, if necessary, take measures.

(*Continued*)

Table 7.3 Design aspects and potential measures for pressure pipelines (*continued*).

Pressure Pipeline Design Aspect	Potential Measures
Can flow be provided for sufficient time in order to break up stagnant air bubbles?	**Yes:** OK **No:** Revise the design.
Is the shortest pump-cycle time long enough for moving a bubble from a declining pipeline?	**Yes:** OK **No:** Increase the pump cycle.
Can the pipeline become empty (mainly for materials with cement lining)?	**Yes:** Design a discharge provision. **No:** OK
Can siphoning occur?	**Yes:** Check operation. **No:** OK

7.2.3 Surge protection design

Table 7.4 Design aspects and potential measures for surge protections against water hammer.

Design Aspect of Surge Protections	Potential Measures
Are the procedures for starting and stopping the system validated?	**Yes:** OK **No:** Check procedures.
Can the installation be operated manually in case of unusual operating situations with the Frequency Converter switched off?	**Yes:** OK **No:** Operational management/revise design
Is the behaviour of the check valve tested in case a surge vessel is necessary?	**Yes:** OK **No:** Check / revise design.
When removing a surge vessel, is one aware of the fact that the air valve is no longer available?	**Yes:** OK **No:** Revise design.

Chapter 8
Commissioning of the system

When commissioning the project, it is crucial to test whether the new or modified system complies with all the design requirements. The acceptance test is actually the baseline-measurement ('benchmark') for later. These measurements are carried out to establish the starting situation of the system (this is also called 'as-built'). The benchmark is essential for detecting deviations later on. Effectively, this means that the agreed-upon performance indicator should be determined. Figure 8.1 shows the process that should be completed.

8.1 BOUNDARY CONDITIONS

Changes made to the system during the period between the design and commissioning might affect the hydraulic capacity. It is, therefore, essential that all the design documents are updated (as-built documentation) and that the consequences of the changes are checked.

Basic assumptions in the dry and wet weather flow rates might also be altered during the period between drafting the design criteria and the commissioning.

It has to be determined which type of performance indicator will be used to evaluate the system.

8.2 DRAFTING AN ACCEPTANCE TEST PLAN

The implemented project might concern only a part of the entire system, for example the renovation of a pumping station. During the acceptance test, however, it is important to examine the entire system, thus the pumping station and the pipeline.

The acceptance plan shall establish which parts of the system are checked and under which circumstances. The acceptance plan must minimally include the following components:

- Water availability
- Leakage-proof and gas-proof pipeline system
- Pumping capacity
- Pipeline capacity
- Dynamic measurements

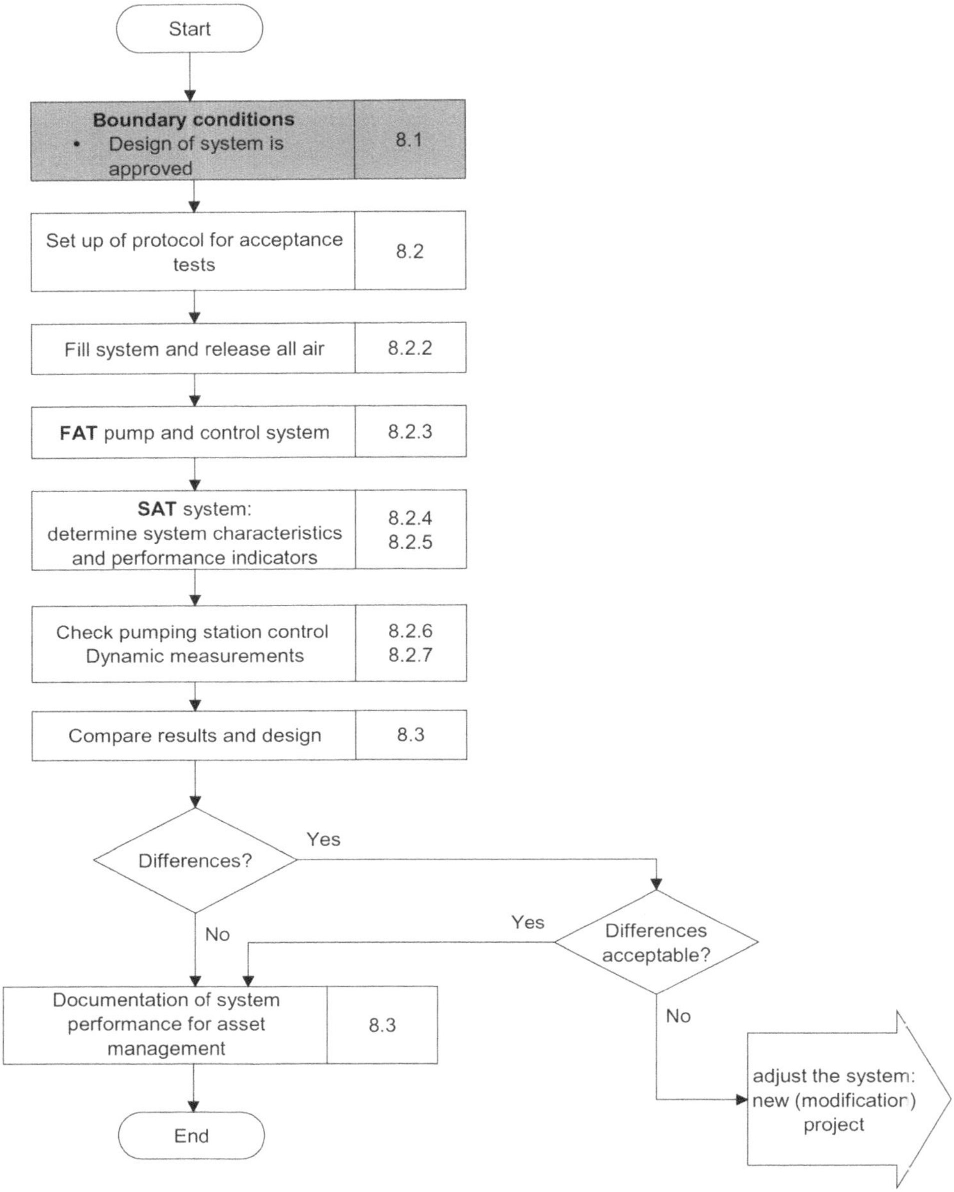

Figure 8.1 Flow chart of the procedures for commissioning the system.

The test must include several minutes of a negative or stationary situation. This means that there is little pressure and flow rate fluctuations and the time signals show a constant behaviour.

During these measurements, we need to observe not only whether the hydraulic operation of the system is correct, but also whether the vibrations and noise levels remain within the acceptable criteria.

8.2.1 Preparations

It is essential however to check the system according to the design criteria, thus the wet weather conditions.

This means that sufficient water must be available in order to carry out a number of relevant measurements. For this purpose, the storage of the sewer system must be used or, alternatively, surface water has to be allowed into the system. If there is storage in the system, the sewer operator must provide the boundary conditions for this:

- What is the volume available in the system?
- Should sub-pumping stations be controlled?
- Criteria for cancelling the test

8.2.2 Leakage-proof and gas-proof pipeline system

The pipeline system has been tested under pressure and must be leakage-proof upon delivery. The contractor must produce registrations that show that during pressurisation, the pressure reduction remained within the determined criteria. All air must be removed from the pipeline system.

Determining whether the system does not contain gas is possible with a simple test, which uses pressure-based storage capacity of the pipeline and compression of the liquid (based on the water hammer theory). This test is described in Appendix B.3.

8.2.3 Pump capacity

The pump manufacturer determines the capacity of the pump in the Factory Acceptance Test (FAT). One or more operating points are indicated in the pump characteristics. This test is not always carried out for standard pumps, yet the pump characteristics must be made available. Each pump in the pumping station must be tested individually. The testers must also make sure that the conditions in which the pump is tested are similar to the conditions in the pumping station. Here, specific attention must be paid to the inflow conditions, the suction pressure and the discharge pressure (the operating point). To determine the capacity of the pump, the following signals must be available:

- Suction level or suction pressure
- Discharge pressure immediately after the pump
- Flow rate
- Pump speed
- Electrical power

The measurements are carried out at the nominal (maximum) speed for at least two flow rates, namely the flow rate at the maximum efficiency point and at $Q = 0$. The latter measuring point is obtained by closing a valve/gate valve at the discharge side.

In a variable speed pump, the operating points are determined by speeds that deliver the minimum capacity. Achieving the lower limit of the pump speed is usually 50% to 60% of the nominal speed.

8.2.4 Check valve

Simultaneously with testing the pump, it is desirable to establish the condition of the check valve in order to be able to register pollutants during the utilisation phase.

The minimum signs that have to be registered are:

- Upstream pressure
- Downstream pressure
- Flow rate
- Opening angle (only possible in rotating types)

In check valves that are equipped with contra-weights or springs, these adjustments must be established.

8.2.5 Pipeline capacity

The pipeline capacity is represented by the system characteristics. To determine these characteristics, the parameters below must be measured (minimally):

- Suction level
- Discharge head or pressure
- Flow rate
- Discharge pressure at the beginning of the pressure pipeline

It is desirable to carry out additional pressure measurements between beginning and end of the pipeline (see also Section 6.2.2). In complex pipelines, the discharge pressure fluctuates due to impact from other pumping stations. During the measurement of capacity, these pumping stations must not operate. The pressure at the beginning of these connecting pipelines can be used for additional pressure registration for the location where these pipelines connect.

8.2.6 Controlling the pumping station

Testing the pumping station will usually be done manually. In pumping stations where several pumps are operated, it is preferable to also evaluate the control of the pumps. As a result of errors in the pump-control programme or the settings, operations could deviate from the manual steering.

To test the complete control regime, a great deal of attention must be devoted to the water supply. This supply must be available for enough time and it must be variable.

An efficient alternative is to link the pumping station operation to simulation software. By testing with 'virtual water', the pumping station operation can be tested by means of a reproducing method.

8.2.7 Dynamic measurements

The dynamic measurements necessary are:

- Testing the effectiveness of the water hammer surge protections
- Determining the dynamic characteristics of the system

The dynamic testing is initiated by an abrupt switching-off of the motor (interrupting the power supply). For the purpose of determining the dynamic characteristics, recording of the pump start is also preferred.

Dynamic measurements cannot be registered by manual readings. Data loggers need to be used in which the time and measuring signals are stored.

Water hammer surge protections

The effectiveness of the water hammer surge protections has to be evaluated for the effect of the surge protection on the system. Evaluating the surge protection itself in fact is only possible by applying surge vessels. For this, prior to the dynamic testing, the following has to be determined:

- Water level in the surge vessel (filling degree)
- Starting pressure of the pressure pipeline
- Flow rate

A time-dependent measurement must be done of the following signals:

- Level in the surge vessel

In order to evaluate the effect of the surge protection, the pressure needs to be evaluated somewhere along the system. If this is missing, then this measurement cannot be used for testing the acceptable pressures in the system. Then, the preferable location derives from the water hammer analysis.

This pressure measurement must occur directly on the relevant pressure pipeline. A measuring point at a connecting pipeline cannot be used. This is because of the propagation of the pressure wave in the system.

Dynamic characteristics

The dynamic increase and decrease of pressure during starting or stopping of the pump is a feature of the system. If something changes in the system, then this dynamic increase and decrease of pressure will also change. That is why it is desirable to establish the increase and decrease of pressure during the acceptance test. The dynamic characteristics will change mainly as a result of gas bubbles in the system.

The dynamic characteristics have no significance if water hammer surge protections are connected to the system. These protections affect the system behaviour. The result is that one cannot assign any significance to the signal.

Dynamic characteristics are established for:

- Pump start
- Pump stop

For details, refer to Appendix B.

8.3 BENCHMARK TESTING

The aforementioned measurements constitute the baseline-measurement ('benchmark') for the system. It is best to carry out each measurement at least twice in order to test the reproducibility. The resulting performance indicators must be tested according to the design.

One should compare the observed capacity of the pumping station and the pipeline with the design basis. If the deviations are great, the measurement should be repeated while devoting extra attention to the correct assumptions:

- Do the instruments work correctly?
- Is all air removed from the system?
- Are all the valves completely open?
- Are the pumping stations that connect to the system shut down?

When the measuring results are reproduced, the results can be properly recorded. In addition, one should record the observed operating points of the pump and the performance indicators.

Deviations with respect to the design must be documented. When accepting the system, the owner/operator accepts these benchmarks as basic assumptions for further capacity-related issues.

If the system is not accepted initially, a problem analysis should follow using additional measurements in order to discover the cause(s). If these steps do not lead to improvement and acceptance of the lower system performance, this should lead to a new or modified design and construction process.

Chapter 9
Maintaining hydraulic capacity

Existing guidelines (including Sewage Guideline, Module C6000 Pumping Station Management) devote a great deal of attention to management and maintenance. Therefore, in this section we want to focus on issues that have not been discussed sufficiently up until now. We will limit ourselves only to the hydraulic aspects of the wastewater transport system. The primary purpose of management is to maintain the design capacity at minimum costs.

The performance of pressure pipeline systems cannot be evaluated by observing only the separate objects. To determine cause and effect, we need to understand the entire system. This might mean that additional disciplines/departments have to be involved in this process.

9.1 MEASUREMENTS IN WASTEWATER TRANSPORT SYSTEMS

Performing measurements in wastewater pipeline systems is a task that every manager faces eventually; whether at the time of delivery or during normal operations, or if there are 'problems' with the system.

This section discusses both on-going monitoring and incidental field measurements, because most of the aspects are vital in both cases.

In all cases, it is necessary to set up a situation tailored to a measurement plan. In a brief specification and in general terms, we will provide some guidance on the content and set-up of a measuring plan.

The purpose of measuring is to collect data and to use this for obtaining information about the performance of the system, in our case a wastewater pressurised pipeline system. An initial vital step, and regretfully often an omitted one, is to formulate precisely which information is necessary to answer the question or issue that has led to initiating the measuring project. The reasoning behind the measuring step has to be formulated in such a way that it enables us to quantify the effectiveness of this step. An example as such might be: "*To be able to detect a trend in the hydraulic resistance of at least 1 metre liquid column (mlc) in a time span of 6 months*".

Based on statistical methods, it can be determined precisely the effectiveness of the measurement set-up (see Schilperoort, 1986).

9.1.1 Parameters and locations

Translating this need for information into requirements for the data should determine what is collected in the next step. The requirements are set in terms of:

- The parameters that have to be measured
- The measuring frequency
- The measuring accuracy
- The measured locations
- The measuring scope per parameter, per location
- Measuring duration (depending on the purpose of measuring, this can vary between 'life-long' and several hours)

Taking into account the geometry and structure of the system that had to be measured, the locations for the measurements must be determined. By using a calculation model, you can examine which places can be considered based on the expected system behaviour.

In existing situations, it is essential to examine whether the selected locations for measuring:

- Are accessible and safe to work in;
- Comply with the built-in requirements of the instrumentation; and
- Contain the necessary provisions (consider for example the availability of electrical power or data communication possibilities).

If the measured location does not comply with the aforementioned requirements, the manager will have to make a choice: Either to adapt the current measuring location or to find an alternative one.

9.1.2 Instrumentation

The selection of the parameters is usually unequivocal and according to the specific issue under study and the system's physics (what is known about it). However, this does not apply for the measuring frequency and the accuracy of the measurements. Both parameters must be determined in coordination with each other and depend mainly on the variability of the measured process.

The measuring frequency must be sufficiently high in order to properly record the fastest processes. A rule of thumb for the accuracy of the measurement is that the order of the largest measured errors must be smaller than the variability of the measured signal on the time scale of the selected measuring frequency. Therefore, measuring frequency and measurement accuracy are highly interrelated.

Schilperoort (1986) has conducted an elaborate study of the relationship between measuring accuracy and measuring frequency with respect to time & space scales of processes.

If during the design phase, one does not have reliable information with regard to the characteristic times of the processes, then it would be sensible to start with as high a measuring frequency as possible. Based on the collected data (up to that point), one can examine whether it is permissible to reduce the frequency.

Determining the duration of the measurement is essential because to a large degree this determines whether structural measures will have to be invested (infrastructure installations, setting up an organisation).

9.1.3 Measuring strategy

Based on the established requirements for frequency, accuracy, locations, parameters and measuring duration, the sensors, the installation method and the method for acquiring data can now be determined. (e.g., manual readings during daily measurements, for long-term measuring trajectories, a wireless data acquisition with online validation). Additionally, it can be determined whether adjustments are still needed in the system in order to enable the measurements. Finally, based on this design, an estimate for the deployment of resources and personnel (e.g. installation, management, data-analysis and reporting) can now be drafted.

Figure 9.1 shows the relationship between the aforementioned topics. However, it should be noted that the effectiveness depends on the quality of the data, in addition to a good set-up of the measurement network. Therefore, missing data or problems with the time synchronization of the various measurement points, which both could have immense negative influences on the effectiveness of the measurement network, can be prevented. Commitment to proper maintenance and regularly conducting a data analysis are effective means for detecting poor data on time.

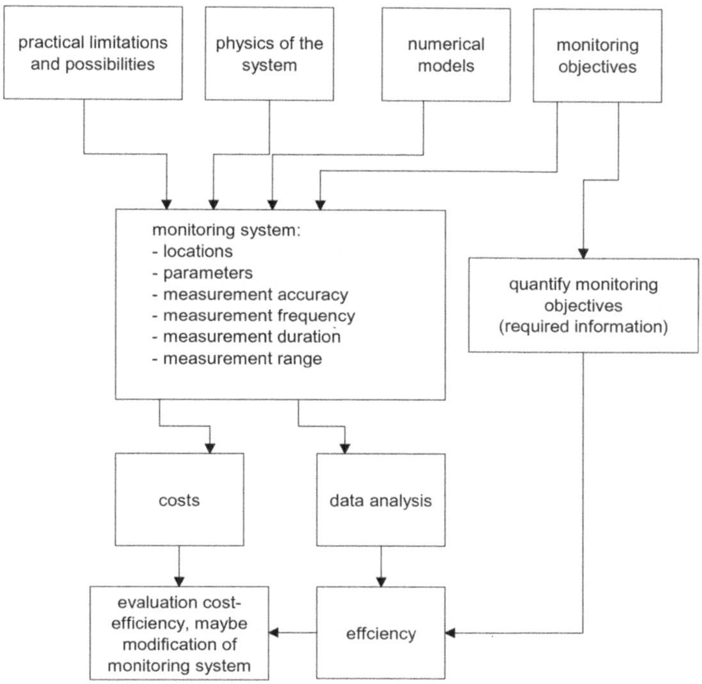

Figure 9.1 The relationship of the monitoring programme.

9.1.4 Implementing field measurements

Carrying out field measurements is time consuming, yet the usefulness of this task should never be underestimated. Comprehensive, procedural preparation, using a procedure manual in which several parties are involved (also from other organisations, such as managers, municipalities, external companies and private parties), costs a great deal of time.

Subjects for the procedure manual are:

- Occupational Health & Safety issues from a technical perspective.
- Availability of sufficient storage in the upstream system.
- Criteria for cancelling the field measurement (precipitation).
- Availability of the right knowledge and skills in order prepare the operational management of the system for field measurement.
- Notifying the responsible organisations and obtaining their permission.

After the procedural preparations, one can begin with the technical preparations. Here too, people from various knowledge disciplines are required.

The technical preparations consist of:

- Selecting location instruments
- Gauges for pressure at high altitudes
- Availability of auxiliary facilities (power, water, data communication)
- Selecting and mounting the instruments
- Adjusting the relevant measuring scope of the instruments
- Calibration of the instruments
- Time synchronisation of the data-acquisition devices
- Availability of sufficient storage in the upstream system
- Dependency on specific weather conditions

Example: When selecting and installing measurement instrumentation, the following knowledge disciplines play a major role: Hydraulics (physical units, measuring scope and measuring frequencies), electro-technical (selection of the right sensor, data acquisition) and the plant operations (building in the instrumentation).

The procedural and technical preparations result in a well-defined measuring plan. This plan contains all the knowledge and details necessary for carrying out a successful field measurement. At the same time, while implementing the field measurement, attention must be paid to the following aspects:

- Dynamic experiments can damage the system. Before carrying out field measurements, it is essential to have proper knowledge of dynamic field measurements. It is highly advisable to perform numerical simulations of every dynamic field measurement in advance in order to get a decisive picture of the expected measurement results. Moreover, the simulations provide an insight into the measurement frequencies, measurement scope and the extreme pressures that might occur elsewhere in the system.
- The effect of wastewater (for example: risk of contamination and toxic H_2S gas).
- For the purpose of measuring, the pumping station control-system is usually adjusted or switched off (manual control).

9.2 CRITERIA FOR CAPACITY REDUCTION

In practice, the concept of a 'good functioning system' is interpreted in different ways (see Section 6.1). The manager usually does not obtain the correct criteria from his organisation to be able to assess objectively whether the wastewater transport system is functioning well. Moreover, 'proper functioning' is often related to the number of disturbances or the number of 'high water' or 'overflow' alerts and not directly to additional energy loss.

To assess the system objectively, the performance indicators were introduced (see Section 6.1). Trends observed in these PIs can quickly indicate a change in the system, which can be addressed in a timely manner. Primarily, this limits the capacity reduction, which prevents the number of overflows and disturbances.

For each system, it has to be established what is the permissible deviation of the PI before taking actions. Three phases are distinguished here:

Phase	Significance
Green	System is working as it should; capacity is close to the benchmark.
Orange	System capacity is decreasing; additional energy reduction; no intervention will lead to additional overflow in the long term.
Red	System capacity is inadequate, great energy loss, immediate danger of overflow.

The organisation's manager must set the transitional criteria between the three phases. On the one hand, this depends on the selected PI, whereas on the other hand, it depends on the impact of the capacity problem on the environment.

If the PI cannot be determined automatically, then it has to be determined periodically and registered. The PI-determination frequency depends on the risk for the pumping station. Gas problems require higher frequencies than scaling, for example (weeks instead of months). Based on experience, the frequency can be adjusted upwards or downwards.

As soon as the system reaches phase orange, the cause of a capacity reduction must be determined through an analysis of the available measurement data. Sometimes, supplementary measurements might be required in order to determine the exact cause of a capacity reduction.

9.3 ANALYSIS OF A CAPACITY PROBLEM

If it transpires that there is a capacity problem, then an analysis must be carried out to determine the cause. The complexity of this analysis depends on the complexity of the system, although most systems can follow a systematic plan to find out what causes the capacity problem.

To eliminate the causes, it is logical to first take the easiest and fastest steps, which can vary per system. In systems that have been operating for years, a good source is the historical knowledge and information. Always consult these sources first.

9.3.1 Localising the problem

Figure 9.2 shows the process for the initial analysis of a capacity problem. The first step for identifying the problem is to make a list of the available measuring data and historical data. It is vital to find out whether the capacity problem occurred spontaneously or whether the energy loss developed into a capacity problem over a longer period.

Specifically in case of a sudden change in the system capacity, first examine whether something has changed in the system. A number of aspects that should be considered are:

- Pollution in the pump impeller
- Erosion of the pump impeller
- Settings of the controls
- Position of the valves

- Static lift of pump
- Changes elsewhere in the system
- Temporary work activities
- Industrial discharges into the sewer system

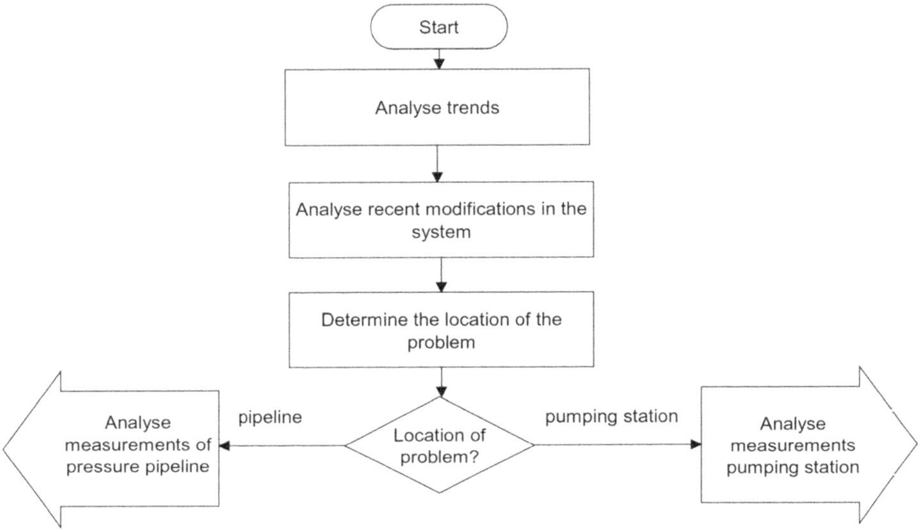

Figure 9.2 Flow chart for the analysis of a capacity problem.

Checking the pumping station log can provide an indication of the problem.

In case of a sudden change in system behaviour, it is also recommended to consult the maintenance logs and reports over the relevant period for all the pumping stations that are part of this system.

Do not forget to include pumping stations that are owned by third parties. Industrial pipelines connected to the system definitely deserve extra attention.

Another aspect that could lead to reduced capacity is an increase in supply. Therefore, check with the sewer manager whether there have been any changes in the supply system.

Based on the correct measuring data, it can be determined where to look for the cause of the capacity problem, i.e. in the pumping station or in the pressure pipelines. Section 6.2 provides more information with regard to measuring volumes and locations.

In order to determine whether the capacity problem occurs in the pumping station or in the pressure pipeline, measurements of the pressure and flow rate at the beginning and pressure at the end of the pipeline are necessary. Subsequently, the resistance characteristics of the pipeline are calculated and these are compared to the design or benchmark. If the resistance characteristics do not deviate significantly from this reference, then the cause of the problem is likely in the pumping station.

Determining the capacity of the pressure pipeline in a multi-branched transport system is not simple because there is no stationary situation due to the numerous pumping stations that have to be switched on and off. In this case, it is sensible to first analyse the pumping station before a great deal of effort is invested in analysing the pipeline.

9.3.2 Analysis of the pumping station

If the first analysis reveals that the cause of the problem is in the pumping station, then further analysis should be carried out (Figure 9.3). Comparing the current measuring data with the data from the acceptance tests (see Chapter 8) or data from the manufacturers will provide a great deal of information.

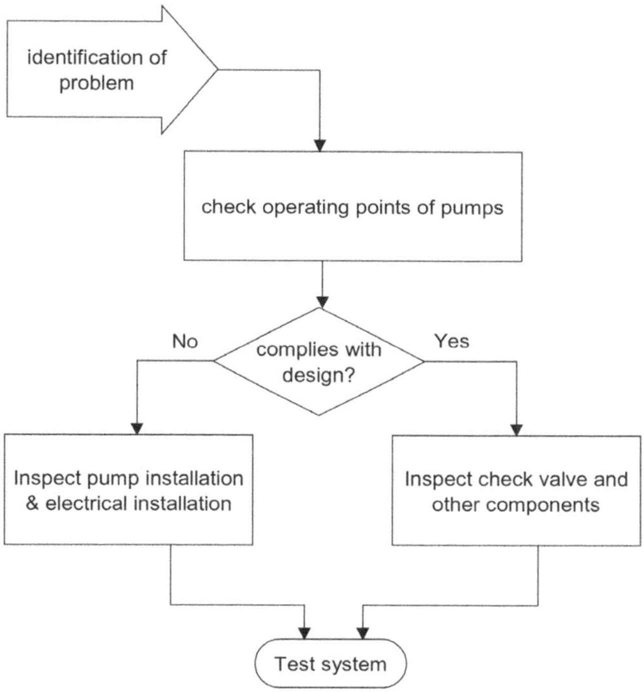

Figure 9.3 Flow chart for the analysis of capacity problems in the pumping station.

First of all, the pumps must be checked. By checking the measured data of the pump (flow rate, lift of pump, number of revolutions, and consumed power) based on the data from the acceptance test, it can be determined whether the pump still complies with the specifications. If this is not the case, the pump impeller must be inspected for pollution and/or wear.

If the analysis of the pump data shows that the pump complies with the specifications, then the source of the problem is usually in the check valve or other appendages in the pumping station. This often calls for a complete inspection of the pumping station.

9.3.3 Analysis of the pressure pipeline

If the system characteristics deviate significantly, then the cause cannot be distinguished easily. The pumphead should be checked first. Predominately in complex pressure pipeline systems the outlet pressure of the pumping station can change due to modifications to other pumping stations in the system.

The next step is an analysis of the dynamic reduction. The increased pipeline resistance can have two causes: 1) Pollution of the pressure pipeline or 2) problematic gas volumes at declining pipeline sections. In practice, gas accumulation is estimated to be the direct or indirect cause of 80% of the capacity problems.

By analysing further, the cause of the capacity reduction can be determined (see Figure 9.4). Here too, first consider the easiest and fastest elimination step. Industrial discharge, for example, might be responsible for grave pollution (scaling) of the pipeline.

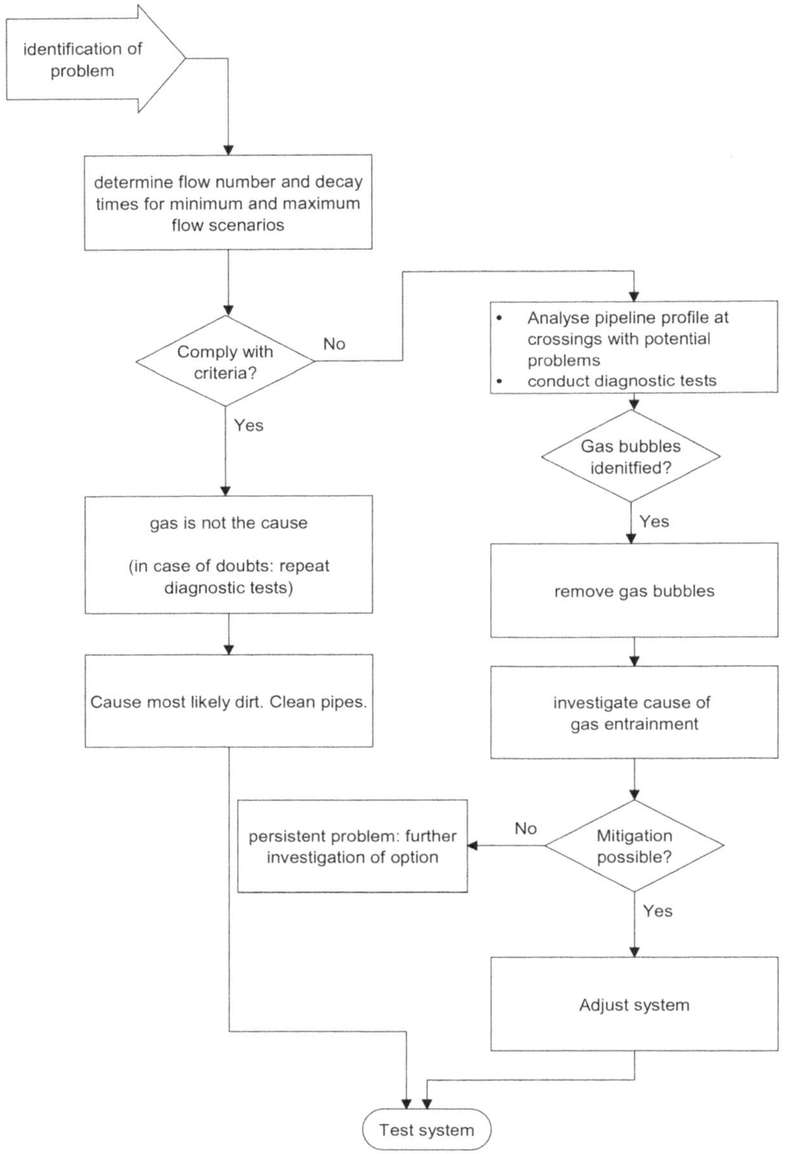

Figure 9.4 Flow chart for the capacity analysis of the pressure pipeline.

Based on the pipeline profile, potential locations (declining pipeline sections) for gas accumulation can be determined. Additionally, examine locations that might have negative pressure because gas tends to escape at low pressures.

Subsequently, based on Section 3.3, for each of these locations the current flow number must be determined. The flow number provides an immediate indication of the chance of gas problems for a particular angle of slope.

If the flow numbers are low enough to allow the gas bubbles to gather in the high points of the pipeline, then a detection measurement must be carried out to determine the presence, location and size of the gas volumes (see Appendix B). This measurement can also be carried out as a confirmation if the analysis shows that the gas bubbles cannot accumulate in the pipeline.

There are three options for removing gas volumes from pipelines:

1. Through air-release valves
2. Streaming
3. Pigging

Which of the three options will be applied initially will depend on the possibilities within the design trajectory. Business-economic aspects related to the operation (time and money) also play a role in the selection and/or order of the options.

The first option, an air-release valve, is the simplest method to implement, but it depends on the presence of suitable air-release valves in the pipeline, access to these valves, and availability of man power.

The volumes of gas can also be removed by means of flow velocity (option 2). This entails streaming during a certain period with a large flow number in order to remove the bubbles from the pipeline. To remove the bubbles effectively, the preferred minimum flow number is $F > 0{,}9$ (*Note*: this depends on the slope angle of the pipe; see Section 3.3; the criterion for gas-volume transport).

If this flow number is not achieved, then the time needed to transport the gas bubble will increase drastically (on the order of hours). Attention also has to be paid to the pipeline profile because gas bubbles are transported from one sag pipe to another. At each declining pipeline, there is the risk that a new gas volume will accumulate.

After emptying out the pipeline, a capacity measurement has to be carried out in order to confirm that the gas has been removed from the pipeline.

A third option is pigging. Pigging with a correct pig offers a guarantee that the pipeline becomes gas-free immediately. A second advantage is that a smaller water buffer is needed while the pipeline is being flushed. Also when pigging, the effectiveness of the cleaning action can be checked by means of a detection method.

In addition to removing the volume of gas from the pipeline, the reason for this gas volume must be examined. By preventing gas in the pressure pipelines, costs can be reduced (both in terms of extra energy reduction and in terms of cleaning activities).

To find the source of air intake, examine the pumping station. Frequent reasons for this include design errors in the air valves of the pumping installation (see Section 4.6), air intake in the receiving basin (see Section 4.5.2) and small leakages where air can enter the pipeline during negative pressure conditions. These minor design errors can usually be remedied with simple adjustments.

If the cause of air intake cannot be discovered in the pumping station, then the entire system will have to be meticulously analysed.

9.3.4 Follow-up maintenance for capacity problems

If the location of the capacity problems is known and a solution is selected and implemented, it is recommended to carry out a new capacity measurement or to analyse the PI frequency in order to test the effectiveness.

It is essential to record in advance the preceding analysis path, its solution and effectiveness so that this information can be drawn upon for solving future capacity problems. The report about the problem can also be instructive for future employees within the organisation.

If capacity reduction is caused by an overly optimistic estimate of preconditions during the design phase, then the transition criteria between the three phases (see Section 9.2), or the design, will have to be adjusted. The basic precepts for the design will also need to be adjusted in order to prevent this problem in future designs.

After a brief period (depending on the cause of the capacity problem), it is desirable to carry out a second control measurement. Another possibility is to analyse the PI frequency during this period. This gives the manager insight into the speed in which the hydraulic capacity decreases and, therefore, the necessary frequency for cleaning. This also serves as a confirmation of both effectiveness of the solution and the analysis of the capacity problem.

The length of time between the two control measurements depends on the cause of the capacity problem. For gas problems, the period is much shorter (in the order of several weeks) than for pollution (in the order of several weeks to months). The primary goal of the second control is to examine whether the analysis of the problem is correct.

If, for example, a capacity problem is incorrectly analysed and pollution of the pressure pipeline is the cause, the solution is pigging. During this step, all the gases are removed from the pipeline. However, since the cause of the accumulated gases is not treated, within a few days or weeks the capacity problem will re-occur. By means of a second measurement (or a new analysis based on the performance indicator), the new capacity problem is determined in time and it can then be concluded that in addition to pollution, gas was also the cause of the problem.

Chapter 10
References

ANSI (1998) American national standard for pump intake design. ANSI-HI-9.8-1998.

Colebrook, C.F. White, C.M. (1937) The reduction of the carrying capacity of pipes with age. Journal of the institute of civil engineers.

Czarnota Z., Geise, M. (2004) Optimal form of wastewater basin of a sewage pumping station (Dutch: Optimale vorm van de vuilwaterkelder van een rioolgemaal). Rioleringswetenschap 14: 79–86.

Deltares (1990–2011) Course on Dynamic Behaviour (Water hammer), Course Series in Hydrodynamics of Pipeline Systems (Dutch: Cursus Dynamisch Gedrag (Waterslag), Cursusreeks Hydrodynamica van Leidingsystemen), Delft, Deltares.

Deltares (1990–2011) Course on Pumps, Course Series in Hydrodynamics of Pipeline Systems (Dutch: Cursus Pompen, Cursusreeks Hydrodynamica van Leidingsystemen, Delft, Deltares.

Deltares (1990–2011) Course on Valves, Course Series in Hydrodynamics of Pipeline Systems (Dutch: Cursus Kleppen, Cursusreeks Hydrodynamica van Leidingsystemen, Delft, Deltares.

Fair G.M (1968) Water purification and waste water treatment and disposal New York: Wiley, 1968.

Hager, W.H. (1999) Cavity outflow from a nearly horizontal pipe. International Journal of Multiphase Flow 25:349–364.

Kranendonk, M. (2007a) Prevention of Air Intake (Dutch: Luchtinslag preventie). Technische Natuurkunde. Rijswijk, TH Rijswijk. Afstudeerverslag.

Kranendonk, M. (2007b) Prevention of Air Intake in Sewage Pumping Stations (Dutch: Preventie van luchtiname door rioolgemalen). Delft, WL | Delft Hydraulics. Meetrapport.

KU Leuven, Laboratorium Hydraulica (2004) Explanation of the Code of Good Practice for Designing Sewage Systems (Dutch: Toelichting bij de Code van goede praktijk voor het ontwerp van rioleringssystemen), KU Leuven.

Lubbers, C. L. (2007) On gas pockets in wastewater pressure mains and their effect on hydraulic performance. Civil engineering and Geosciences. Delft, Delft University of Technology. PhD thesis, 2007.

NEN (1997) NEN-EN 1671:1997, Pressure sewage systems outside buildings (Dutch: Overdrukrioleringen buiten gebouwen).

NEN (2008) NEN-EN 752:2008, Drain and sewer systems outside buildings (Dutch: Buitenriolering).

NEN (2011) NEN-EN 3650-1:2011, Requirements for Pipeline Systems – Part 1: General (Dutch: Eisen voor buisleidingsystemen – Deel 1: Algemeen).

Pothof, I.W.M. (2011) Co-current air-water flow in downward sloping pipes; transport of capacity reducing gas pocket in wastewater mains, PhD thesis, Delft University of Technology, Delft.

Pothof, I., Clemens, F. (2010) On elongated air pockets in downward sloping and inclined pipes, J. Hydr. Research 48(4), pp. 499–503. http://dx.doi.org/10.1080/00221686.2010.491651

Pothof, I.W.M., Clemens, F.H.L.R. (2011) Experimental study of air-water flow in downward sloping pipes, Int. J. Multiphase Flow 37, 278–292. http://dx.doi.org/10.1016/j.ijmultiphaseflow.2010.10.006

Pothof, I.W.M., Kooij, C., Clemens, F.H.L.R. (2011a) Seven years of research in the joint industry project CAPWAT, Proceedings 12th Int. Conf. Urban Drainage (ICUD), 11–16 September 2011, Porto Alegre, pp. 1–9.

Pothof, I.W.M., Tukker, M., Kooij, C., Clemens, F.H.L.R. (2011b) Guidelines on capacity reducing gas pockets in wastewater mains, Proceedings Computing and Control for the Water Industry 2011 (CCWI), 5–7 September 2011, Exeter, pp. 1–6.

Prosser, M.J. (1977) The hydraulic design of pump sumps and intakes. Cranfield, Bedford. BHRA, 1977.

Rajaratnam, N. and Muralidhar, D. (1964) End depth for circular channels. ASCE Journal of Hydraulic Engineering, Vol. 90.

Schilperoort, T. (1986) General considerations on hydrological networks. Design Aspects of Hydrological Networks. TNO Committee on Hydrological Research Publ. 35: 5–13.

Smit. A. (2006) Air entrainment with plunging jets. Civil engineering and Geosciences. Delft, Delft University of Technology. Master thesis.

Tukker, M. J. (2007) Head Loss in Downward Pipes Caused by Gas Pockets (Dutch: Energieverlies in dalende leidingen ten gevolge van gasbellen). Delft, Deltares | Delft Hydraulics, 2007

Wickenhäuser, M. (2008) Two-phase flow in air-venting systems of pressurized mine galleries (German: Zweiphasenströmung in Entlüftungssystemen von Drukstollen) (in German) Miteilungen 205, Versuchsanstalt für Wasserbau, Hydrologie und Glaziologie (VAW), H.-E Minor, Hrsg. ETH Zürich.

Wijdieks, J. (1978) Water hammer phenomena in Liquid-transport Pipes (Dutch: Waterslagverschijnselen in Vloeistoftransportleidingen). Symposium 'Waterslag in PVC Leidingen'. Waterloopkundig Laboratorium, Delft.

Appendix A

Theoretical background

It is important to be familiar with certain characteristics of liquids, basic equations and energy losses when calculating liquid movements in closed pipelines. In this chapter, relevant theoretical knowledge is provided to the user of this manual. Further information can be found in scientific literature and textbooks on pipeline hydraulics. The reference list at the end of the main text will also serve as guidance.

A.1 CHARACTERISTICS OF DRINKING WATER

The characteristics of drinking water are presented in Table A.1. Note that compressibility modulus and surface tension can vary significantly for wastewater.

Table A.1 Characteristics of water.

Characteristic	Value
Density	998 kg/m^3
Dynamic viscosity	1.0 Pa s or Ns/m^2
Kinematic viscosity	1.0×10^{-6} m^2/s
Compressibility modulus	2.1×10^9 N/m^2
Vapour pressure (at 20°C)	0.017 bara
Surface tension (clean water)	0.072 N/m

A.1.1 Density

The density (ρ_L) of a liquid is identified as the mass per unit of volume [kg/m^3]. The density depends on the temperature and pressure, but is independent of the gravitational acceleration (g, [m/s^2]).

A.1.2 Viscosity

The dynamic viscosity η of a liquid is a measure for the shear stress τ that liquid layers that slide over each other in a layered or laminar flow exercise on one another [kg/ms] or [Ns/m^2]. The shear stress of a Newton liquid can be presented as follows:

$$\tau = \eta \cdot \frac{v_1 - v_2}{z} \tag{A.1}$$

The shear stress is proportional to the average flow velocity and the viscosity of the liquid (Figure A.1). The dynamic viscosity depends on the temperature. The higher the temperature the smaller η is and therefore, the smaller the shear stresses are.

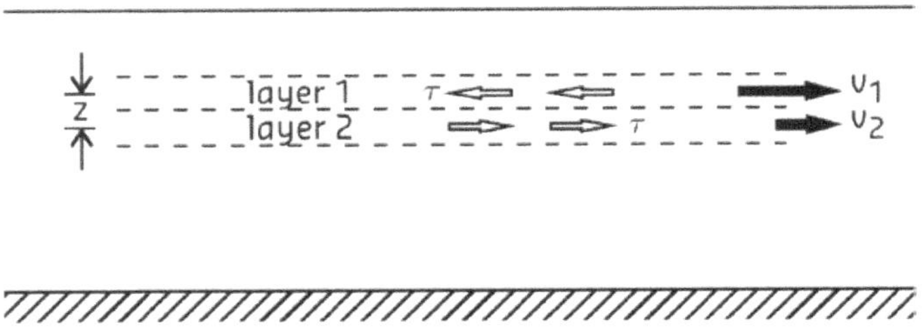

Figure A.1 Resistance between liquid layers.

If two layers in a liquid are at an infinitely small distance from each other (*dz)* and the velocity difference between the two layers is *dv*, then the shear stress can be presented as follows:

$$\tau = \eta \frac{dv}{dz} \tag{A.2}$$

in which:

*d*v = Change in velocity
dz = Change in depth
dv/*dz* = Velocity gradient

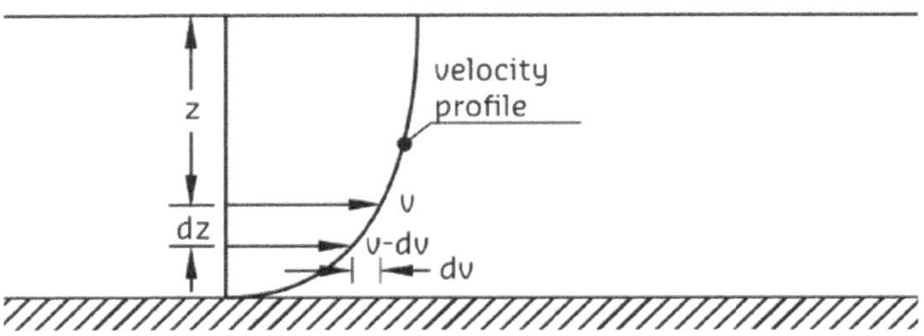

Figure A.2 Velocity gradient.

The kinematic viscosity ν [m^2/s] is often used in the fluid mechanics instead of the dynamic viscosity, η:

$$\nu = \frac{\eta}{\rho} \tag{A.3}$$

A.1.3 Compressibility modulus

The compressibility of the liquid is determined by the compressibility modulus, *K*.

$$dP = -K\frac{dV}{V} \tag{A.4}$$

Equation (A.4) shows the relation between pressure increase and volume change.

A.1.4 Surface tension

Surface tension (σ [N/m]) is defined as a force that works on a boundary surface between gas and a liquid. This force is in balance with the resistance of the boundary surface against deformation (Figure A.3). The relationship between the pressure difference and the surface tension is shown in Equation (A.5). This stress is mostly determined by the composition of the wastewater, the temperature and the velocity in which the boundary surface is created.

$$\Delta p = \frac{2\sigma}{r} \tag{A.5}$$

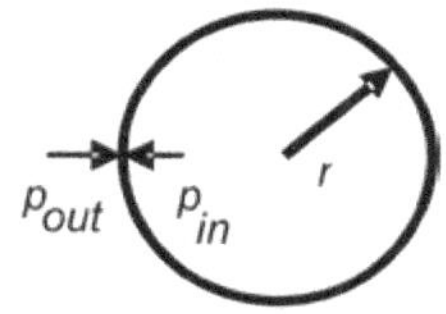

Figure A.3 Pressure forces on a drop of water.

The H_2O molecules are kept together in water by the strong hydrogen bonds. In bulk, these forces will be equally strong in all directions and neutralise each other. This interaction is much weaker in the boundary surface due to the presence of neighbouring molecules. As a result, the molecules on the surface contain more energy.

Fats and soap have a hydrophilic part that dissolves easily in water and a hydrophobic part that does not dissolve easily in water. The hydrophobic part wants to exit the water matrix and the pollution will move with a certain speed (molecular diffusion) to the boundary surface. At the boundary surface, the forces between the water molecules will continue to weaken and as a result the surface tension will decrease. An increasing concentration of pollution will further decrease the surface tension until the entire boundary layer is filled. Subsequently, the lowest surface tension is measured.

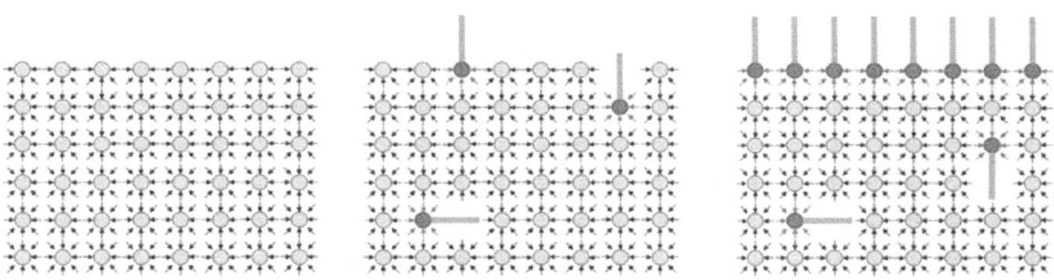

Figure A.4 Pollution with a hydrophobic part moving to the boundary surface (top).

When measuring the static surface tension, a boundary layer is created and with time the pollution moves to the boundary layer.

In case of the dynamic measurement with various speeds, a small gas bubble is created on a capillary in the liquid. If a gas pocket is created quickly, the pollutions in the liquid will not have a chance to reach the created boundary layer. The measured surface tension will then be practically equal to that of clear liquid. The molecular diffusion of the pollution determines the speed at which it affects the surface tension. Cleaning products, such as soaps, are made to work fast and consist of relatively small movable molecules that affect the surface tension in less than a second. Fats and proteins are usually larger cumbersome molecules and their effect is only noticeable after 5 to 10 seconds.

A.1.5 Vapour pressure

The vapour pressure (P_v) is the value of the absolute pressure in which the liquid vaporises [Pa] or [bara]. The vapour pressure rises with increasing temperature. The vapour pressure of water at 100°C is 1.03 bara; the vapour pressure at 10°C is only 0.012 bara. Vaporisation that occurs due to an increase of the temperature at constant pressure is called cooking. Vaporisation that occurs due to a decrease of pressure at constant temperature is called cavitation. Cavitation occurs when the absolute pressure in a drinking water pipeline with water of 10°C decreases to 0.012 bara (almost vacuum).

A.2 BASIC EQUATIONS

A.2.1 Reynolds number

Reynolds showed that there are two types of flows: laminar or layered flows and turbulent flows. The types of flows depend on the flow velocity, v [m/s], and the kinematic viscosity, ν [m²/s]. The higher the flow speed, v, and the less viscous the liquid, the more turbulent the flow is. Reynolds presented the connection between these two quantities in the dimensionless Reynolds number Re [-]:

$$Re = \frac{v.L_K}{\nu} \tag{A.6}$$

in which, L_K [m] is the characteristic length with regard to the geometry of the considered pipeline. As long as the velocity is low, this is called layered or laminar flow. The kinematic viscosity of the liquid ensures that disturbances of the flow are rectified. The first swirls in laminar flow occur where the velocity gradient $\frac{dv}{dz}$ is high and the stabilising effect of the wall does not dominate at a certain distance from the wall. The swirls will fill the entire flow pattern with the exception of the laminar boundary layer on the wall. The flow has then become turbulent.

The diameter, D [m], is used as characteristic length for completely filled round pipes and the hydraulic radius R [m] is used for open channels, canals and rivers. The latter is equivalent to the cross-sectional area, A [m²], through which the water flows, divided by the wet circumference, O [m] (contact length with bottom and wall):

$$R = \frac{A}{O} \tag{A.7}$$

For a completely filled pipe as well as a half-filled pipe, $R = \frac{1}{4}D$ applies. From this, the Reynolds number follows for filled or partially filled pipe:

$$\mathrm{Re} = \frac{4 \cdot v \cdot R}{\nu} \tag{A.8}$$

Nikuradse (1933) showed that the transition area from laminar to turbulent flow begins at Reynolds numbers larger than 2320. A flow for which Re < 2320 is always laminar. A flow with 2320 < Re < 4000 can be laminar or turbulent, depending on the previous history of the flow. If Re > 4000, the flow is always turbulent.

A.2.2 Bernoulli's principle

Bernoulli's principle derives from the law of conservation of (mechanical) energy:

$$E_{tot} = E_{pot} + E_{kin} = mgh + \tfrac{1}{2}\, mv^2 = \text{const.} \tag{A.9}$$

Dividing by *mg* gives:

$$E = h_1 + \frac{v_1^2}{2g} + z_1 = h_2 + \frac{v_2^2}{2g} + z_2 \tag{A.10}$$

in which:

E	Energy head of the liquid (constant)	[m]
$h_i = \frac{p_i}{\rho g}$	Pressure head (water column in a riser pipe placed in that point)	[m]
Z	Elevation of the point (distance between the point under consideration and the reference surface)	[m]
$h + z$	Piezometric level with regard to the selected reference surface	[m]
$\frac{v^2}{2g}$	velocity head	[m]

According to Bernoulli's principle, in an ideal liquid, the sum of the piezometric head and the velocity head along a streamline is equal to the energy head. The line that connects all the energy levels of a certain streamline is called the energy line. The line that is drawn through the piezometric levels is the pressure line.

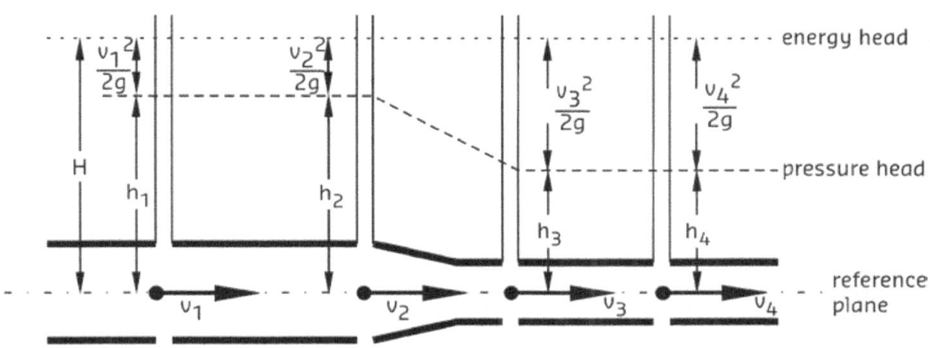

Figure A.5 Energy head and pressure head.

Bernoulli's principle only applies if:

- All the points under consideration lie on one streamline.
- No energy is added or withdrawn from the flow between the points.
- The flow velocity in a given point is consistent (stationary flow).

When energy supply or withdrawal does occur, for example as a result of friction, then equation (A.10) changes into:

$$h_1 + \frac{v_1^2}{2g} + z_1 = h_2 + \frac{v_2^2}{2g} + z_2 + \Delta H_{1,2} \quad \text{(A.11)}$$

In which ΔH [m] is the difference in energy head.

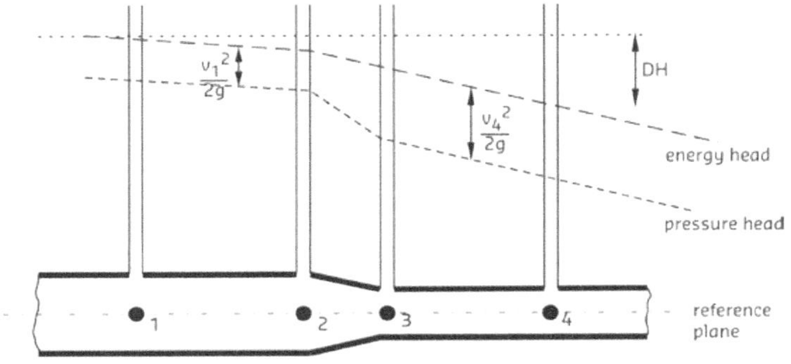

Figure A.6 Lines of energy and pressure levels in case of friction.

A.3 TRANSPORT LOSSES

Transporting a liquid through a pipeline costs energy. The following equation can recap these energy losses:

$$\Delta H = \xi \cdot \frac{v^2}{2g} \quad \text{(A.12)}$$

in which ξ [-] is a coefficient of losses that should be determined, depending on the cause of the loss. In principle, the aforementioned equation claims that all the losses are proportional to the velocity head $\frac{v^2}{2g}$ of the liquid. In addition, a distinction can be made between local losses and friction losses.

A.3.1 Pipe friction losses

Resistance (the shear stress along the pipe wall) that the flowing liquid undergoes causes friction. This is expressed in a pressure difference across the pipeline. The Darcy-Weisbach equation provides the pressure difference, ΔH (i.e. pressure level difference expressed in metres):

$$\Delta H = \lambda \frac{L}{D} \frac{v^2}{2g} \quad \text{(A.13)}$$

Or, expressed in flow rate:

$$\Delta H = \frac{\lambda L}{D}\frac{\left(\frac{4Q}{\pi D^2}\right)^2}{2g} = \frac{8\lambda L}{\pi^2 g D^5}Q^2 \tag{A.14}$$

The friction coefficient λ depends on the characteristics of the liquid and the characteristics of the pipe. The biggest influence on this coefficient comes from the equivalent wall roughness *k*.

The equivalent wall roughness was originally introduced by Nikuradse (1933), who carried out measurements on pipelines covered with sand grains of the same size (evenly distributed inside the pipeline). The pipeline diameter as well as the diameter of the sand grains varied.

Subsequently, Colebrook and White (Colebrook, 1937) carried out experiments. They used commercially available pipes in which the roughness was unequally divided over the pipe wall. Colebrook and White provided the following equation for the relationship between the Darcy frictional factor and the wall roughness:

$$\frac{1}{\sqrt{\lambda}} = -2\log\left(\frac{2{,}51}{\mathrm{Re}\cdot\sqrt{\lambda}} + \frac{k}{3{,}71\cdot D}\right) \tag{A.15}$$

Generally, a wall roughness k of 0.5–1 mm is used when designing wastewater transport pipelines. Other values for *k* can be found in Table A.2.

Table A.2 Wall roughness for various materials.

Material	Value of *k* in mm
Plastic pipes with a small diameter	
Asphalted asbestos-cement pipes	0.01
Centrifugally asphalted, centrifugal cast-iron and seamless steel pipes	
Synthetic coating	
Asphalted drawn steel pipes	0.02
Smooth asbestos-cement pipes	
New seamless steel pipes	
Pre-stressed concrete according to Freyssinet	
Asphalted, centrifugal cast-iron and welded steel pipes	0.05
PVC pipes with a large diameter	
Asbestos-cement pipes	
Cracked cement	
Mains according to the German Association of the Gas & Water industry (German: DVGW)	0.1
New galvanised steel pipes	
Asphalted cast-iron pipes	
Slightly rusted seamless steel pipes	
New welded steel pipes	
Glazed vitrified clay pipes	0.02
New centrifugal cast-iron pipes	
Pre-stressed concrete	
Smooth sewers – straight and without connecting pipes	

Moderately rusted seamless steel pipes	
Centrifugal cement coating	
Smooth sewers with connecting pipes	
Distribution pipelines according to the German Association of the Gas & Water industry (German: DVGW)	0.05
Centrifugal concrete pipes	
New cast-iron pipes	
Smooth concrete	
Moderately rusted welded steel pipes	
Concrete pipes	
Smooth drain pipes	
Concrete	
Straight sewers without connecting pipes	1
New riveted steel pipes	
Highly rusted seamless steel pipes	
Cast-iron pipes with slight pock formation	
Sewers with connecting pipes	
Highly rusted welded steel pipes	
Porous concrete drain pipelines	
New riveted steel pipes with overlapping part	2
Rough concrete	
Highly rusted riveted steel pipes	
Cast-iron pipes with high pock formation	
Highly rusted riveted steel pipes with overlapping part	5
Cast-iron pipes with very high pock formation	

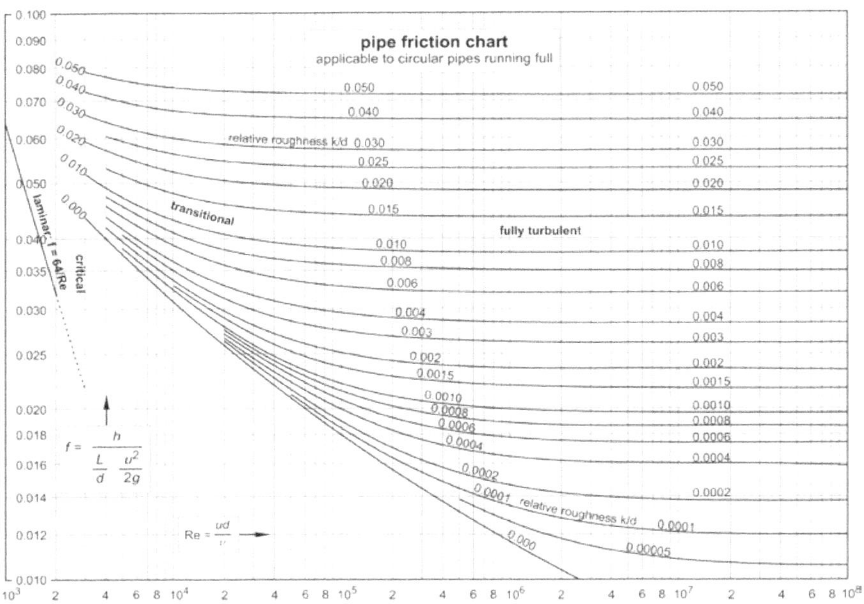

Figure A.7 Moody diagram.

Equation (A.15) is an implicit equation: the friction coefficient λ appears on the left side *and* on the right side and can be solved iteratively by computers. Alternatively, the Moody diagram is a useful tool: The friction coefficient, λ, can be found for the entire flow area – from laminar to turbulent. The friction coefficient, λ, can be found after calculating the Reynolds number and determining the relative wall roughness, $\frac{k}{D}$.

A.3.2 Local losses

The local losses are the result of the fact that the kinematic energy cannot be turned entirely into potential energy in case of a decrease of the flow speed. Part of the kinematic energy is necessary for propelling swirls and vortexes that are created because streamlines are no longer able to run parallel to the wall.

As stated before, local losses can be calculated with Equation (A.12). Depending on the cause of the local loss, we can determine the coefficient ξ that belongs to the characteristic loss situations in a pipeline system described below.

Inlet losses

At the inlet, flow contraction (μ) occurs and creates loss. This contraction (constriction of the liquid jet) leads to a flow resistance.

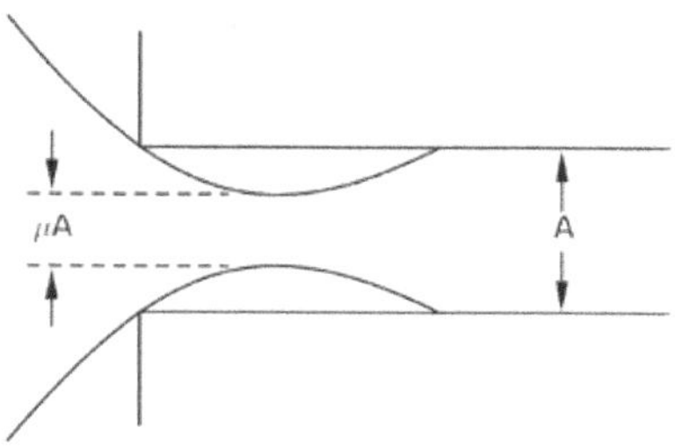

Figure A.8 Contraction.

The contraction coefficient m is determined by the rounding off the inlet opening and it has the following relationship with ξ_i:

$$\mu = \frac{1}{1+\sqrt{\xi_i}} \rightarrow \xi_i = \left(\frac{1}{\mu} - 1\right)^2 \quad \text{(A.16)}$$

Table A.3 Coefficient ξ_i for various constructions.

Type of Inlet	Sketch of the Construction	ξ_i	μ
Sharp corners	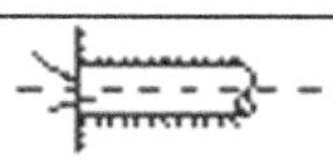	0.4 … 0.5	0.58…0.61
Protruding inwards	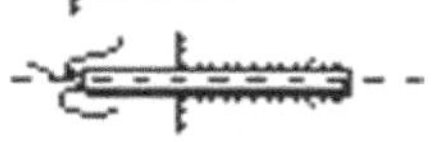	0.8 … 1.0	0.5 … 0.53

Type of Inlet	**Sketch of the Construction**	ξ_i	μ
Bevelled corners	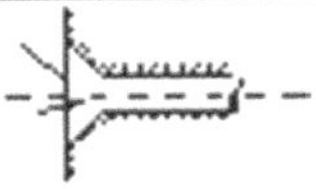	0.2 … 0.3	0.65…0.69
Rounding with a small radius	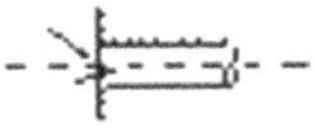	0.1	0.75
Rounding with a large radius	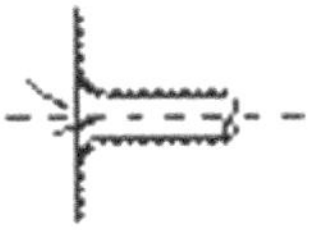	0	1
Sharp, with angle α	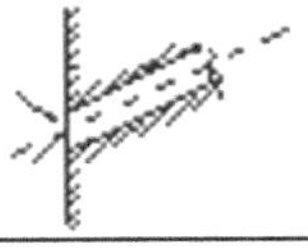	$0.5 + 0.32 \cos\alpha + 0.2 \cos^2\alpha$	

Outlet losses

In case of unhindered discharge of a liquid into a large open storage, the entire velocity head is ultimately lost because the velocity becomes zero after a certain distance from the outlet. The outlet loss can be reduced by gradual or phased expansion. The ratio of the surface of the flow pipe, A_1 [m^2], and the surface of the profile after discharge, A_2 [m^2], determine the level of the outlet loss coefficient, x_u:

$$\xi_u = \left(1 - \frac{A_1}{A_2}\right)^2 \tag{A.17}$$

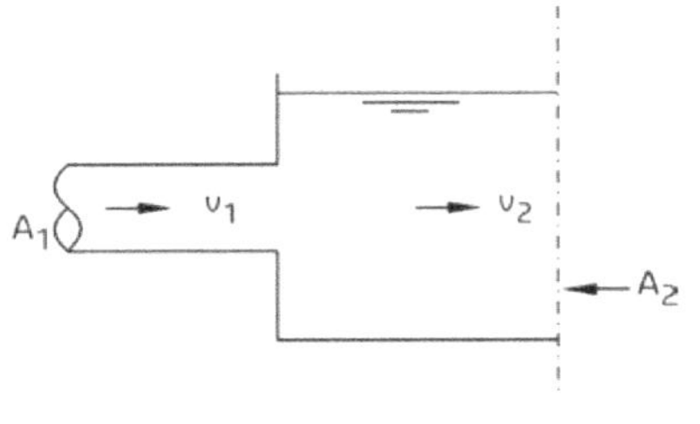

Figure A.9 Ratio of the surfaces.

Bend losses

According to Idelchik (2007), the coefficient of bend losses is a function of the ratio bend radius to pipe diameter, wall roughness, Reynolds number, bend angle and cross-section type. The following generic relation is used:

$$\xi = \xi_{loc} + \xi_{fr} \tag{A.18}$$

in which

$$\xi_{loc} = C_1 \cdot C_2 \cdot C_3 \tag{A.19}$$

and the friction coefficient, ξ_{fr}, being a function of bend ratio, pipe diameter, wall roughness and Reynolds number. Idelchik (2007) provides detailed formulae for these three coefficients. Due to their complexity, they are not repeated here. Alternatively, one can use design graphs. Miller (1990) uses a similar approach with correction coefficients but proposes different equations to account for the above parameters.

The following table presents a few key values and should be seen as rules of thumb. For design purposes, accurate friction coefficients shall be made.

Table A.4 Coefficient ξ for bend losses in a circular pipeline.

$\frac{r}{D}$	0.5	1	2	5
ξ	2	0.3	0.15	0.1

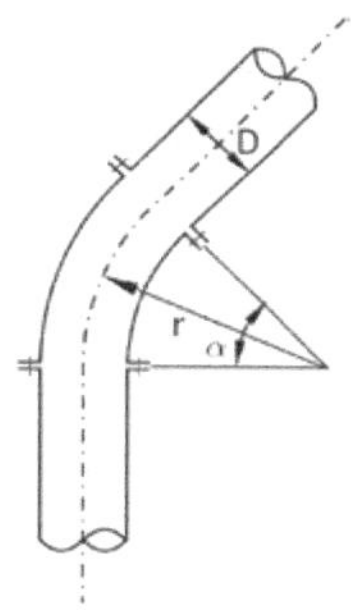

Figure A.10 Definition of the bend radius r, pipe diameter D and angle α.

According to Idelchik (2007), bend losses in circular pipelines with $\frac{r}{D} < 3$ can be calculated using $\xi = A \cdot B$ in which A and B are defined as follows:

A = 0.9 sin(α) if the angle α = ≤70°
A = 1 if the angle α = 90°
A = 0.7 + 0.35 (α /90) if the angle α = ≥100°

and

$\mathbf{B} = 0.21 \cdot \left(\frac{r}{D}\right)^{-2,5}$ if $\frac{r}{D} = 0.5 - 1.0$
$\mathbf{B} = 0.21 \cdot \left(\frac{r}{D}\right)^{-0,5}$ if $\frac{r}{D} => 1.0$

Bend losses in circular pipelines with $\frac{r}{D} \geq 3$ can be calculated using $\xi = 0,0175 \cdot A \cdot \alpha \cdot \frac{r}{D}$ [(A.20)], in which A is defined as follows:

$A = \frac{20}{Re^{0.65}}\left(\frac{D}{2r}\right)^{0.75}$	if $50 < Re \cdot \sqrt{\frac{D}{2r}} < 600$
$A = \frac{10.4}{Re^{0.55}}\left(\frac{D}{2r}\right)^{0.255}$	if $600 < Re \cdot \sqrt{\frac{D}{2r}} < 1400$
$A = \frac{5}{Re^{0.45}}\left(\frac{D}{2r}\right)^{0.275}$	if $1400 < Re \cdot \sqrt{\frac{D}{2r}} < 5000$

Elbow losses

Idelchik (2007) provides the most comprehensive literature on elbow losses including all major parameters of influence. A very simple (and, therefore, understandable) expression is given by Hütte (1919), who describes the coefficient for elbow losses depending on the elbow angle. The following applies to a pipeline with a diameter of at least 30 mm:

$$\xi = \sin(0.5 \cdot \alpha)^2 + 2 \cdot \sin(0.5 \cdot \alpha)^4 \quad \text{(A.21)}$$

If the diameter becomes smaller, then ξ becomes larger.

Table A.5 Coefficients ξ for elbow losses for a pipeline with diameter of 30 mm or larger.

α	5°	10°	15°	20°	30°	40°	45°	60°	80°	90°	100°	120°	140°	160°
ξ_k	0.00	0.01	0.02	0.03	0.08	0.14	0.19	0.38	0.75	1.00	1.28	1.88	2.44	2.85

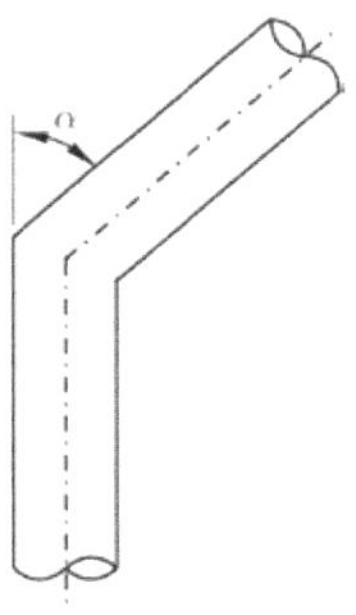

Figure A.11 Definition of the elbow angle.

Note that this is a first order approximation. More complex formulae (e.g. Idelchik, 2007) will give results that are more accurate.

Losses due to contraction and expansion of the pipe

The losses due to contraction and expansion of the pipeline depend on the cross-sectional area. The following applies to an abrupt expansion:

$$\xi_{V_2} = \left(\frac{A_2}{A_1} - 1\right)^2 \quad \text{(A.22)}$$

$$\xi_{v_1} = \left(\frac{A_1}{A_2}\right)^2 \left(\frac{A_2}{A_1} - 1\right)^2 = \left(1 - \frac{A_1}{A_2}\right)^2 \qquad \text{(A.23)}$$

The outlet loss is $\Delta H_v = \xi_{v_1} \cdot \frac{v_1^2}{2g}$ or $\Delta H_v = \xi_{v_2} \cdot \frac{v_2^2}{2g}$.

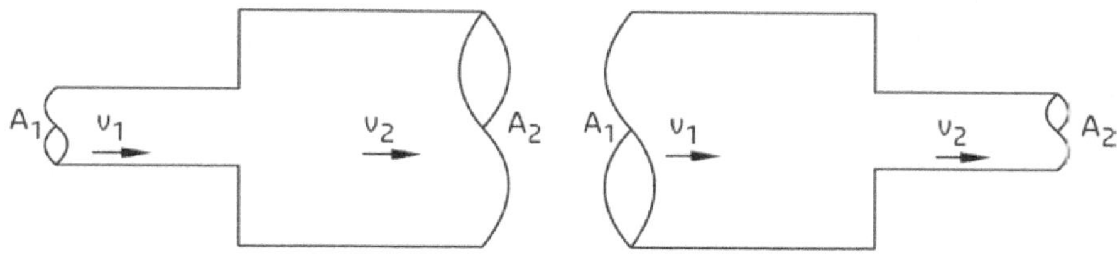

Figure A.12 Cross-sectional areas.

An abrupt contraction is comparable to an inlet loss. The loss is thus mostly determined by the geometry of the contraction, which leads to a certain contraction loss coefficient, μ.

$$\xi_2 = \left(\frac{1}{\mu} - 1\right)^2 \qquad \text{(A.24)}$$

A.4 RECEIVING BASINS AND INLET STRUCTURES

A.4.1 Rules of thumb for the design of receiving basins

Many standard solutions are available for pump intakes as has been summarised by Prosser (1977) in 'The hydraulic design of pumps sumps and intakes and the Hydraulic Institute Standards for centrifugal and rotary pumps', for example.

The requirements for a pump sump are:

- Symmetric basin shape for uniform pump approach flow
- No dead corners to avoid sedimentation
- Correct position of the suction mouth in relation to the bottom and walls

The following rules of thumb for the dimensions of pump compartments and the location of the pump apply to vertical submerged pumps that are placed in pump compartments. These are based on the ANSI/HI 9.8 1998 standard, which has been mainly drafted for clean water. Note that no additional guidelines exist for specific wastewater applications and problems. This can be seen, for example, in the requirement for low approach flow velocities (clean water), while increased flow velocities are preferred to transport wastewater (sedimentation prevention).

The distances and sizes are made dimensionless by expressing them in multiples of the bell mouth diameter.

A.4.1.1 Bottom clearance

The recommended bottom clearance of the pump is between 0.3 and 0.5 times the bell mouth diameter. Higher clearance can result in an unstable flow to the impeller. Lower clearance leads to higher velocities below the bell mouth. This can lead to separation of flow of the bell mouth, which results in a non-uniform velocity profile at the entry of the impeller.

A.4.1.2 Backwall distance

The recommended distance between the pump shaft and the rear wall is 0.75 times the diameter of the bell mouth. Based on practical experience, Deltares often uses a distance between 0.75 and 1 times the diameter of the bell mouth. A relative distance larger than 1 can lead to stagnant water behind the pump. Stagnant water is unstable and can lead to the creation of vortices and tornados as well as to settlement of waste. A smaller distance can lead to poor feeding to the impeller at the side of the backwall. This can lead to a non-uniform velocity profile at the entry of the impeller.

A.4.1.3 Width of the pump compartment

The geometry and dimensions of the pump compartments must provide a stable and uniform feed from all directions to the bell mouth of the pump. The width of the pump compartment greatly influences the suction behaviour of the pump. The recommended width of the pump compartment is twice the bell mouth diameter of the pump. A smaller width results in higher flow velocities in the direction of the pump, which can lead to higher swirl and the creation of vortex. The obstruction of the pump housing becomes relatively big so that the water flows less easily around the pump in order to feed properly from the backside.

A.4.1.4 Length of the pump compartment

The approach flow of the pump should be straight from the front and not under an angle. A certain straight length before the pump must guarantee this approaching flow. Generally, a length of five-times the suction mouth diameter or more is used upstream of the pumps. Often, there is a distribution chamber and filter compartments if a pump basin contains several pumps. This may cause strong swirling flow. In such a case, a length of six-times the bell mouth diameter is used in order to rectify the oblique approach flow.

A.4.1.5 Inclined walls (side walls/floor)

If there are inclined walls in the pump basin (for example, an ascending basin floor), then the angle should be less than 10 degrees. Larger angles mean a greater risk for separation of flow, creation of vortices and instability in the approach flow.

A.4.1.6 Submergence depth

If a pump is directly connected to a suction basin, then the bell mouth of the pump must be sufficiently submerged below the water level to prevent air entrainment.

The required submergence depth that is necessary to prevent air entrainment is a function of the Froude number, based on the velocity and diameter of the bell mouth. The Froude number is defined as follows:

$$Fr_D = \frac{V_D}{\sqrt{gD}} \tag{A.25}$$

in which:

Fr_D = Froude based on the suction mouth of the pump	[-]
V_D = Velocity at the beginning of the suction mouth	[m/s]
g = Gravitational acceleration	[9.81 m/s^2]
D = bell mouth diameter	[m]

Reference material provides correlations for determining the necessary submerge depth. One of these correlations derives from the ANSI/HI 9.8 1998 norm for scale modelling of pump basins. This standard is accepted worldwide for testing pump basins in a scale model. The rule of thumb for the required submergence is based on a wide range of experimental pump basin data. It states the following:

$$S_{ANSI}/D = 1 + 2.3Fr_D \tag{A.26}$$

The aforementioned rule of thumb is a guideline, which means that a certain tolerance exists. This depends on the pump basin design and the specific approach flow. Therefore, a certain range is sometimes indicated around the ANSI submergence depth. Deltares uses a range based on literature, acquired knowledge and experience from pump basin projects, which is indicated as S_{upper} and S_{lower}:

$$S_{lower}/D = 1 + 2Fr_D \tag{A.27}$$

$$S_{upper}/D = 1.5 + 2.5Fr_D \tag{A.28}$$

If one assumes that the flow rate remains the same, the critical submergence depth becomes larger when a smaller bell mouth diameter is chosen. Table A.6 provides an example for the required submergence depth of the bell mouth of 2 pumps with different bell mouth diameters. Table A.6 shows that the inlet velocity, in relation to the bell mouth diameter of pump 1, is relatively higher than that of pump 2. This results in a higher Froude number for pump 1. As a result, there is a higher critical submerge depth for pump 1 with a smaller suction mouth diameter.

Table A.6 Required submergence depth for pumps in pump basins.

		Inlet of Bell Mouth		Submergence Depth		
Pump	**Flow rate [m^3/h]**	**Diameter [mm]**	**Velocity [m/s]**	**S_{lower} [m]**	**S_{ANSI} [m]**	**S_{upper} [m]**
1	15.530	1.300	3.25	2.8	3.1	3.8
2	15.530	1.600	2.14	2.1	2.3	2.9

To illustrate this:

Figure A.13 shows experimental data as well as the above rule of thumb as a function of the flow velocity in the suction pipe and water depth. This shows that the required submergence depth increases in case of higher suction velocities. The required submergence depth is the lowest for a width of 2D and a bottom clearance of 0.5 D (X/D = 0.5 and Y/D = 0.5 in Figure A.14).

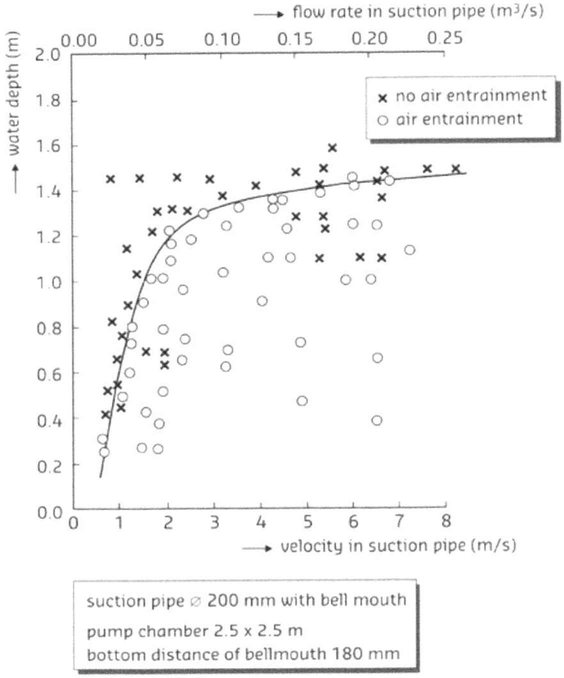

Figure A.13 Results of a measurement to determine the correlation between the necessary water depth and the suction pump flow rate (Wijdieks, 1965).

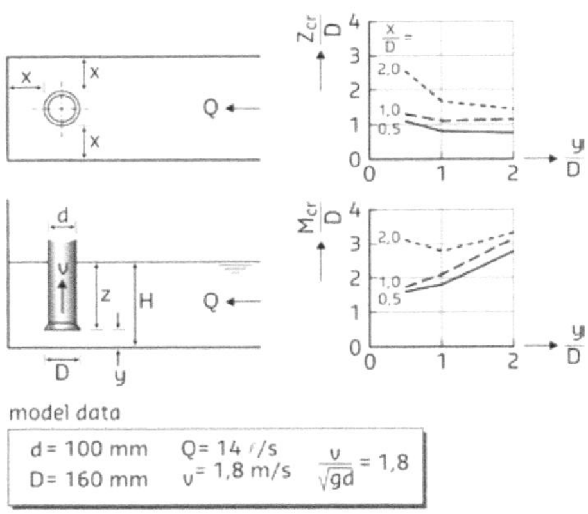

Figure A.14 Required water coverage and the total water depth for preventing air suction in a straight approach channel as a function of the location and size of the bell mouth (Wijdieks, 1965).

The critical water coverage, under which air entrainment occurs, appears to depend on the approach flow. presents experimental results that show that the required submergence depth drastically increases the more the approaching flow becomes asymmetrical (b/B becomes smaller). Figure A.16 shows that the position of the bell mouth also influences the critical water coverage. In the design of the tested pump compartment, it appeared that the position of the bell mouth right along the walls of the compartment is the most advantageous (mostly uniform approaching flow). The most disadvantageous location is the centre of the pump compartment, because the flow swirls around this centre. These experimental results show that we must strive for an approaching flow of the pump that is as uniform as possible.

A.4.1.7 Summary of the rules of thumb

The table below provides a summary of the aforementioned rules of thumb for designing the pump compartment.

	Recommended Dimensions Expressed in Bell Mouth Diameter (D)
Clearance to the bottom	0.3–0.5 D
Distance between pump shaft and rear wall	0.75–1.0 D
Width of the pump compartment	2.0 D
Length of the pump compartment	>5D
Inclined walls	<10°
Submergence depth	D(1+2.3Fr)

Examples of pump sump layouts are provided in Figure A.17 and Figure A.18. The dimensions of the basin are expressed in the diameter of the bell mouth. Such set-ups work satisfactorily when the velocity distribution at a distance of 3 D upstream of the pump is uniform and the average velocity does not exceed 0.3 m/s. (Note that this is based on experimental results and design guidelines for clean water. Higher velocities are often chosen for wastewater to avoid excessive sedimentation.)

Choosing a suitable layout becomes more difficult when there are several pumps (Figure A.18).

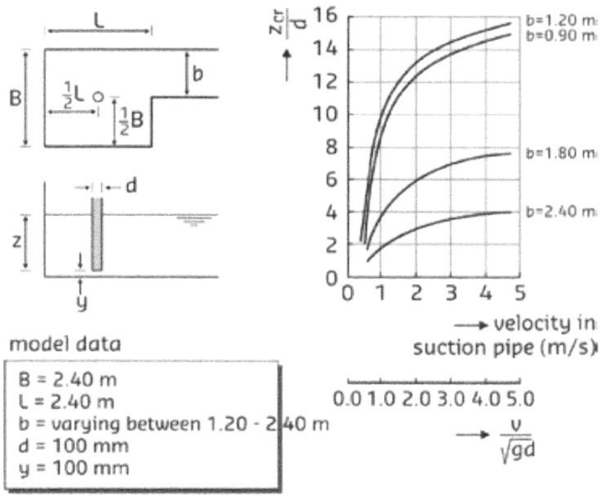

Figure A.15 Influence of pump compartment design and approach flow channel on critical submergence depth (= required water coverage for preventing air suction), Wijdieks (1965).

As a result of lack of space, desired restrictions of the civil engineering construction, desired enlargement of capacities, it is often impossible to comply with the above rules for ideal approach flow. Yet, a good approach flow can often be realised by additional measures. However, these measures need to be verified by means of a physical model study.

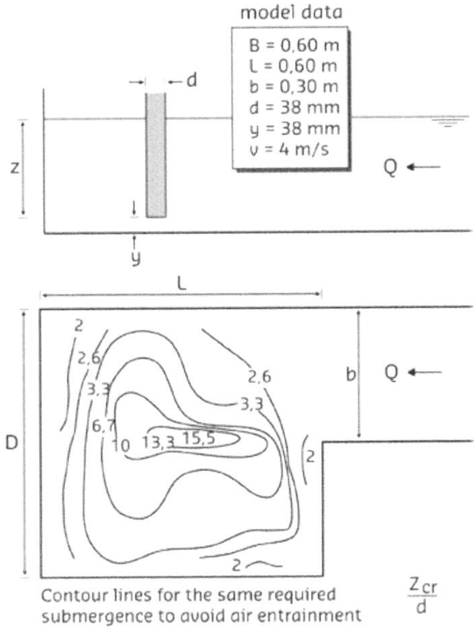

Figure A.16 Influence of the location of the suction pipe in the pump compartment (Wijdieks, 1965).

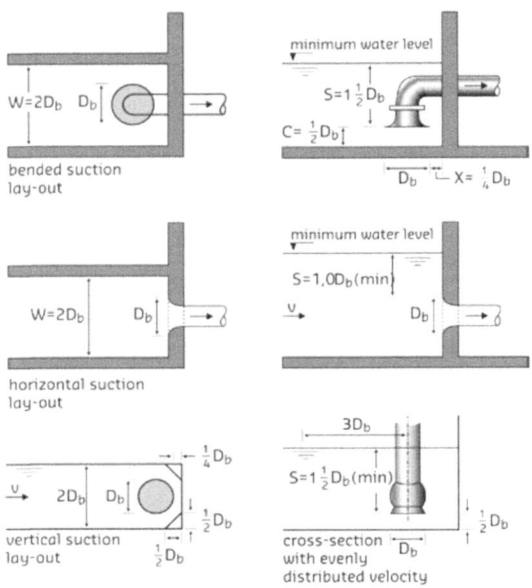

Figure A.17 Set-up of one pump in a pump compartment.

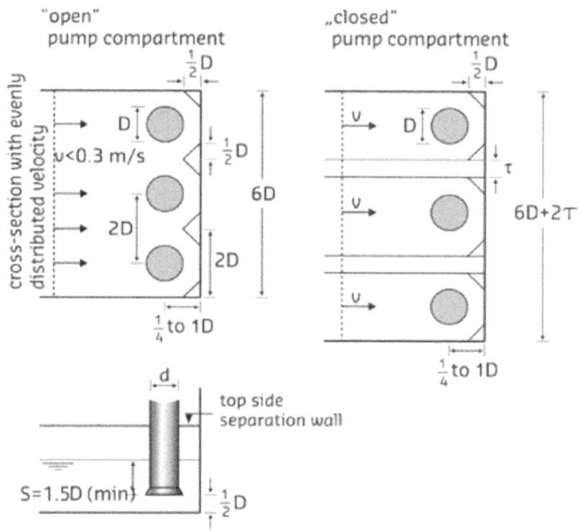

Figure A.18 Set-up of several pumps in a pump compartment.

A.4.2 Air entrainment in pump basins

In addition to the behaviour of gas pockets in declining pipe sections, the CAPWAT research (2003–2009) also studied how the gas enters the pipeline. The most probable cause is air entrainment in the pump basins (see Figure A.19).

Figure A.19 Photo of the test set-up of Kranendonk (2007a). The air pockets are sucked into the pump.

Smit's research (2007) provided much knowledge about these phenomena. Kranendonk (2007a) and Kranendonk (2007b) studied various measures to limit air entrainment or to prevent it. This section first summarises the results from Smit's research in order to familiarise the reader with the phenomena and subsequently, presents a few constructions for limiting air entrainment.

Air entrainment in pump basins is caused by a water flow that falls from a certain height on the free water level of the pump basin. The falling water (the water jet) entrains air in the free water level of the basin and subsequently pulls this air to a certain water depth.

The following parameters are important for designing pump basins:

- The horizontal distance until the end of the plume of bubbles
- The maximum depth to which the plume of bubbles reaches.

Both parameters depend on the water flow rate and the geometry of the pump basin.

The vertical distance depends on the flow velocity v_e of the water at the end of the incoming sewer. This flow velocity is determined by the flow rate, Q_w and the surface area, A_e, through which the water flows.

$$v_e = \frac{Q_w}{A_e} \tag{A.29}$$

The flow velocity, as well as the surface through which the water flows, are unknown parameters. Smit (2007) also measured the water depth at the point of outflow (the final depth h_e). The results correlated very well with the results of Rajaratnam & Muralidkar (1964) who created the following equation:

$$\frac{Q_w}{\sqrt{g}D^{2,5}} = 1.54\left(\frac{h_e}{D}\right)^{\frac{11}{6}} \tag{A.30}$$

Now that the final depth h_e is known, the surface A_e through which the water flows is also known and the final flow velocity v_e can therefore be calculated.

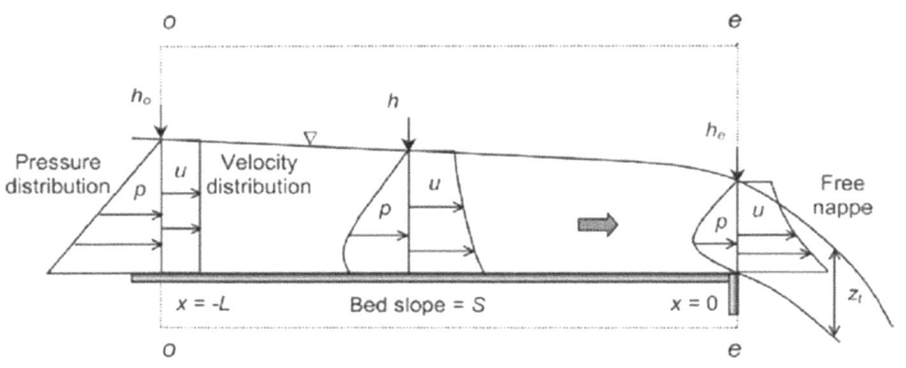

Figure A.20 Schematic presentation of the free outflow and the hydraulic parameters.

The trajectory of the falling water jet is described by the following equations:

$$\left(x(t);y(t)\right) = \left(v_e t; H_e + w_e t - \tfrac{1}{2}gt^2\right) \tag{A.31}$$

in which:

$$H_e = H + \tfrac{2}{3}h_e \tag{A.32}$$

In other words, the mass centre of the flow is added to the outflow height H.

The vertical flow velocity of the water plays a role in determining the trajectory, because the flow is curved at the point of outflow. However, this vertical flow velocity is unknown. By ignoring the vertical flow velocity, we reach a conservative estimate of the horizontal falling distance of the water jet:

$$L_{jx} = v_e \sqrt{\frac{2H_e}{g}} \tag{A.33}$$

The length of the bubble plume is also measured by Smit (2007) and resulted in the following rule of thumb with L_a [m] and Q_w [m^3/h].

$$L_a = \sqrt{2{,}5 \cdot 10^{-4} \cdot Q_w} \tag{A.34}$$

The actual air inlet volume by the pump is very difficult to determine. Smit (2007) as well as Kranendonk (2007b) carried out measurements of the air intake, but the results of these measurements showed strong dependency on the quality of the water, pump flow rate and basin geometry.

A.4.3 Measures against air entrainment

Air entrainment in pump basins can be prevented. During the CAPWAT study (phase II, 2006–2009) various measures for minimising air entrainment by the pump were examined by Kranendonk (2007b). In this section, two of the solutions are discussed: the deflection plate and the fall pipe.

The deflection plate is a plate that is placed before the point of outflow.

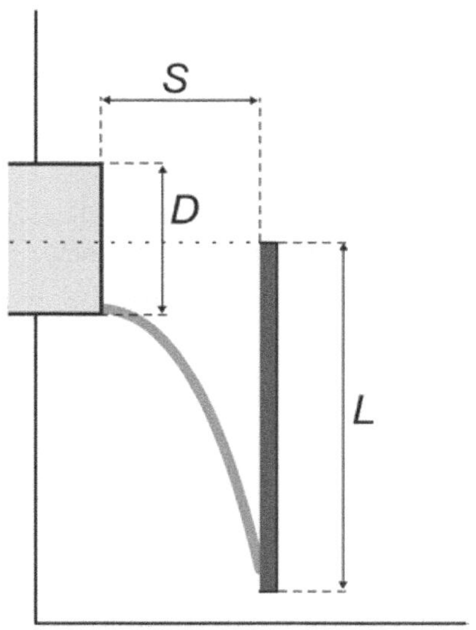

Figure A.21 Dimensions of the deflection plate.

The top of the plate is placed halfway the incoming pipeline, so that any blockage from debris or waste can be removed.

The deflection plate can continue up to the bottom of the receiving basin, but this is not always necessary. The minimum length of the deflection plate depends on the flow velocity and construction aspects of the plate:

$$L = \tfrac{1}{2}D + \tfrac{1}{2}g\left(\frac{S}{V_e}\right)^2 \tag{A.35}$$

The outflow velocity, v_e, must be based on the dry weather supply flow rate (Q_{dwf} in m^3/s) and can be specified by:

$$v_e \approx \left(\frac{Q_{dwf}\, g}{0.25\pi D}\right)^{1/3} = 2.32\left(\frac{Q_{dwf}}{D}\right)^{1/3} \quad \text{(A.36)}$$

in which the outflow depth is equal to:

$$y_e \approx \left(\frac{Q_{dwf}}{0.25\pi D\sqrt{g}}\right)^{2/3} = 0.55\left(\frac{Q_{dwf}}{D}\right)^{2/3} \quad \text{(A.37)}$$

The distance between the pipe and the plate (*S*) must be at least equal to the diameter of the pipe in order to prevent stagnation. The width of the plate must also be equal or larger than the diameter of the pipe.

In practice, it will be easier to build the deflection plate up to the bottom of the receiving basin. Then, the bottom supports the plate so that only the horizontal forces have to be absorbed. The water will flow around the sides of the deflection plate.

Horizontal forces caused by the water jet from the incoming sewer can be estimated as follows:

$$\sum F_D = \frac{1}{2} \cdot C_D \cdot \rho_l \cdot v_e^2 \cdot A \quad \text{(A.38)}$$

The highest possible wet weather supply flow rate must be used for the flow rate. For the drag coefficient of flat plates perpendicular to the flow direction $C_D = 2$ can be assumed. For a flat plate, this value is independent of the Reynolds number. Note that the hydraulic design engineer has to think carefully about the project-specific values for A (cross-section) and v_e (velocity).

Figure A.22 shows how the deflection plate works. The water jet is deflected and the air bubbles rise to the water surface before they reach the pump.

Figure A.22 Photo of test set-up with deflection plate: 1D width, 1D from the point of outflow (Kranendonk (2007a)).

The fall pipe is a pipe that is mounted perpendicular on the incoming sewer by means of a T-piece. The diameter of the fall pipe must be equal to the diameter of the incoming pipeline. The distance between the bottom of the basin and the fall pipe must be at least 1D (see Figure A.23).

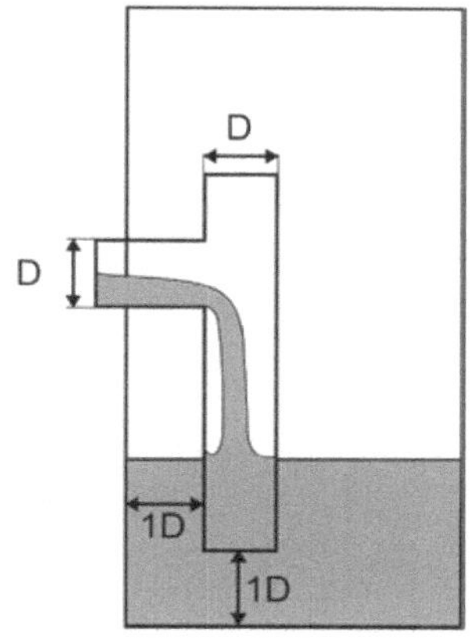

Figure A.23 Dimensions of the fall pipe.

The T-piece at the top of the pipe allows free in and outflow of air. This way, water can flow out of the pipe from the bottom. When exiting the fall pipe, the air bubbles do not have velocity in a horizontal direction, which is why they float back upwards along the outside of the fall pipe (see Figure A.24).

Figure A.24 Fall pipe in the test set-up (Kranendonk (2007a)).

Another solution used in practise is covering the outflow opening with a rubber flap. The advantage of this is that an obstacle can be easily removed from the sewer.

A.5. PUMPS IN PIPELINE SYSTEMS

A.5.1 System characteristics and operating point

To explain a number of basic definitions, a simple system is considered consisting of a single pump and a single pipeline. The pump transports liquid to a point at a higher location. In addition to the static pump head H_p, the pumps must also overcome the resistance that the flow encounters Δh_{dyn}, the dynamic pump head. The static pump head is the difference between the outlet level and the suction level. The outlet level is the pressure level at the outlet side of the pump without flow rate by the pump. The suction level is the pressure level at the suction side of the pump without flow rate by the pump. The dynamic pump head depends on the flow rate.

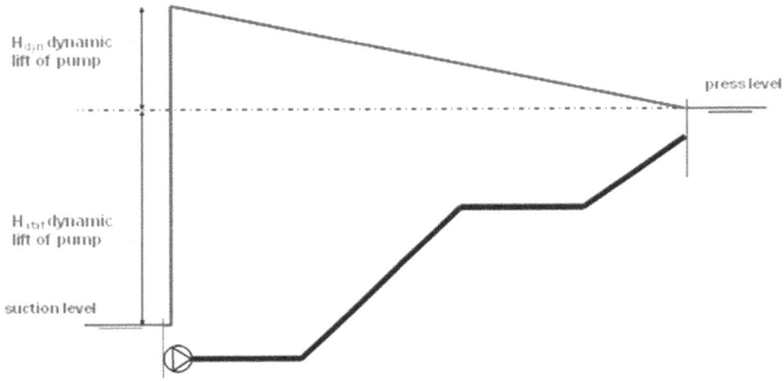

Figure A.25 Definition of suction level, outlet level, static pump head and dynamic pump head.

The pump curve H=f(Q) with various system characteristics of the pipeline is drawn in Figure A.26. The total pump head that the pump must overcome is $H_t = H_p + \Delta h_{dyn}$. The operating point of the pump is determined by the intersection with the system resistance curve in point C. The pump head that the pump can produce at this flow rate corresponds with the total pump head that is necessary to transport this flow rate through the system.

When the resistance increases ($H_t' = f(Q)$), for example as a result of partial clogging or air accumulation, it intersects the pump curve at point C'. The produced pump head increases to H_t' and the produced flow rate decreases to Q'.

A.5.2 Changes in the operating point due to fluctuation in the static pump head

In practice, the static pump head of the system might change, for example because a tank is filled. The dynamic pump head does not change as a result.

Figure A.27 shows how the system characteristics change. The total pump head goes from H1 to H2 and the flow rate decreases from Q1 to Q2. The line between C1 and C2 is the operating range of the pump.

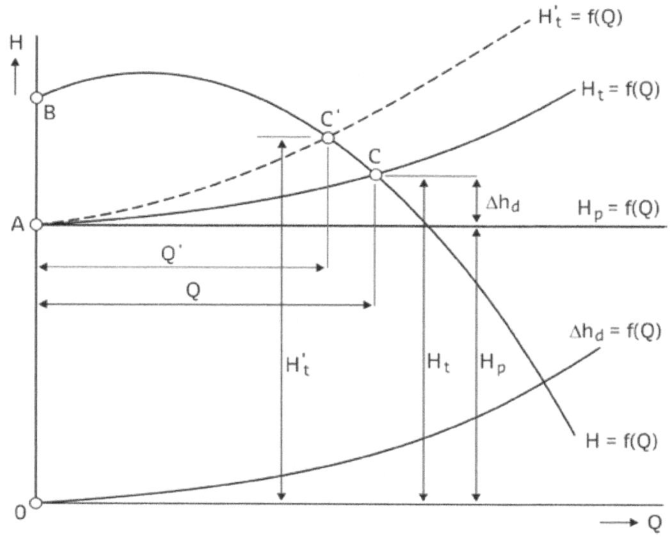

Figure A.26 Pump curve and system characteristics of a pipeline system.

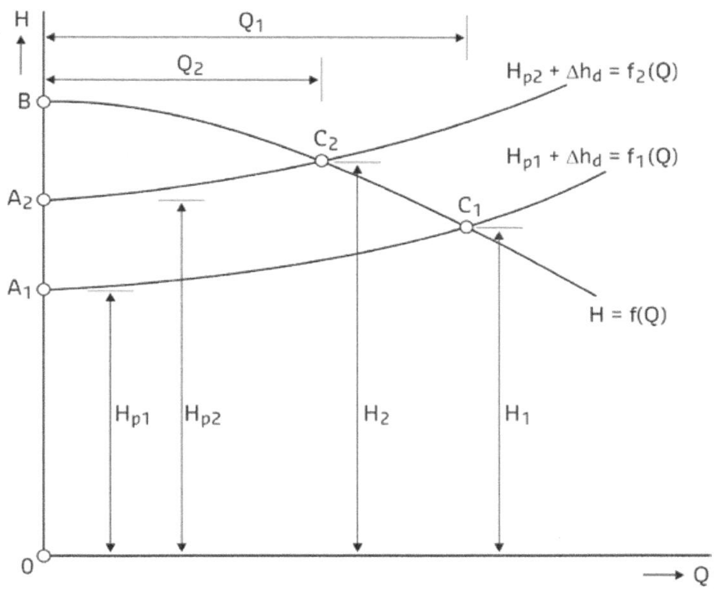

Figure A.27 Pump curve and system characteristics with fluctuating static pump head.

A.5.3 Changes in the operating point due to fluctuating transport losses

Figure A.28 shows what happens to the system curve when the wall roughness increases.

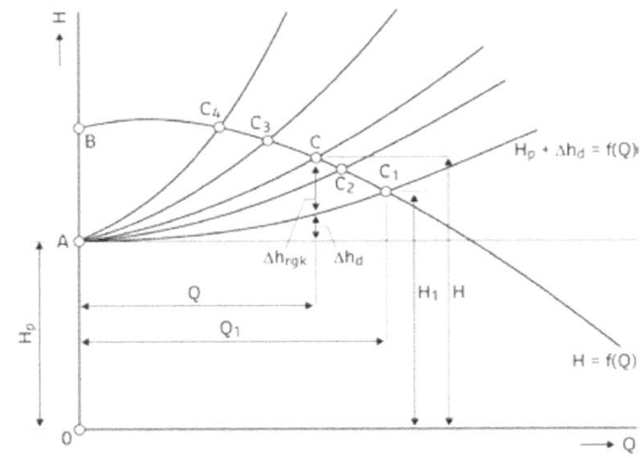

Figure A.28 Pump curve and system characteristics in a pipeline system with varying transport losses.

A.5.4 Comparing various pumps

If the system characteristics change, the type of pump greatly determines the behaviour of the entire system in terms of the operating point. Figure A.29 shows a steep and a flat pump characteristic. The static pump head varies between $H_{p,min}$ and $H_{p,max}$. For the pump with a steep pump curve, this produces an operational range between C_4 and C_5 with a flow rate between Q_4 and Q_5. For the pump with a flat pump curve, this produces an operating range between C_1 and C_3 with a flow rate between Q_1 and Q_3.

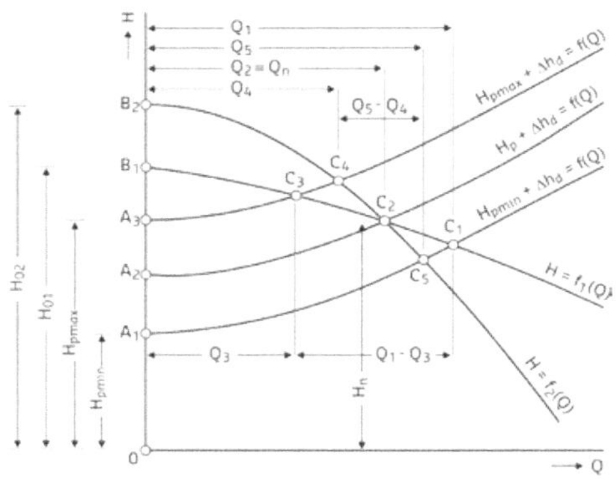

Figure A.29 Comparing various pump curves.

Pumps with a flat characteristic produce relatively smaller changes in the total pump head and big changes in the flow rate. This can be very useful when the system must handle a big difference in flow rate.

Pumps with a steep characteristic produce smaller changes in flow rate with relatively greater variety in the pump head. This can be useful when a greater variety in flow rate is undesired.

A.5.5 Parallel and serial operation of pumps

Parallel and serial pumps are applied when a single pump does not achieve the required flow rate and/or the required pump head.

Two identical pumps that are parallel connected generate a total pump characteristic of which the total pump head is equal to the individual pump but of which the flow rate is theoretically twice as high. Figure A.30 shows the Q-H curve $H = f_1(Q)$ of the individual pump and the combined Q-H curve $H = f_2(Q)$ of the two parallel-connected pumps.

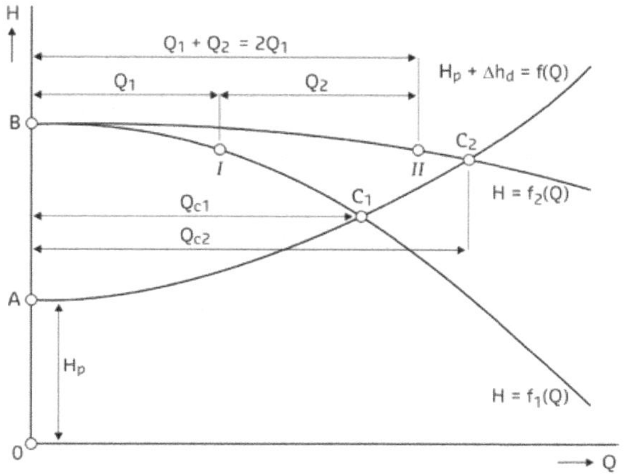

Figure A.30 Parallel operation of two pumps.

The operating point of the system is the intersection C_2 of the combined Q-H curve and the system characteristic. It is important to know that the two pumps that work parallel always produce in reality less flow rate than twice the flow rate of one pump connected to the same system. This derives from Figure A.30, which shows that $Q_{c2} < 2 \cdot Q_{c1}$. This is because the hydraulic resistance increases to the power of 2 of the flow rate.

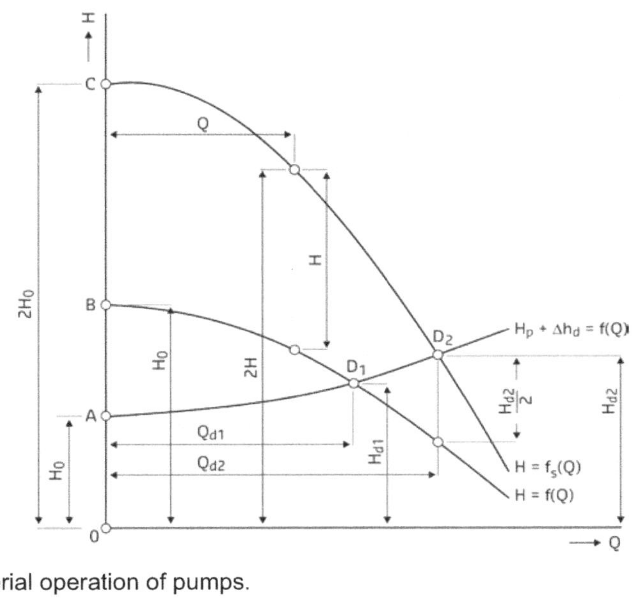

Figure A.31 Serial operation of pumps.

Two identical pumps that are connected in series produce a total pump characteristic of which the total pump head is twice as high as that of the individual pump but in which the flow rate remains the same. When two pumps are connected in series, then the pumps produce the same flow rate per definition. The total pump head is twice as high as that of the individual pump head (with identical pumps) (see Figure A.31).

A.5.6 Changes in pump speed

When the pump characteristics of a pump at a certain pump speed are known, the affinity rules determine how these will change when the number of revolutions changes. Accordingly, the hydraulic characteristics of two identical pumps operating with different pump speeds can be deduced from each other by means of affinity rules. This can be applied to pumps with the same design and construction and it can be used to determine the pump characteristics for different pump speeds n.

When ignoring the viscosity effects, the affinity rules for the pump flow rate (Q), pump head (H) and hydraulic power (P) are as follows:

$$\frac{Q_b}{Q_a} = \frac{n_b}{n_a} \tag{A.39}$$

$$\frac{H_b}{H_a} = \left(\frac{n_b}{n_a}\right)^2 \tag{A.40}$$

$$\frac{P_b}{P_a} = \left(\frac{n_b}{n_a}\right)^3 \tag{A.41}$$

A.6 VALVES

A.6.1 Shut-off valves

The principle of influencing the flow rate by a valve is based on a change of the cross-sectional area in the valve housing by mounting an obstacle (restriction) in which the size of the restriction is a function of the valve position. The use of such a restriction in a pipeline (Figure A.32) provides insight into the interdependencies among various hydraulic parameters.

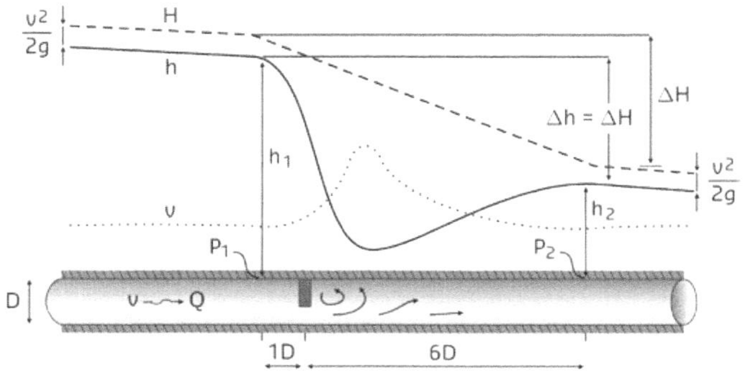

Figure A.32 Pressure and velocity curve with control valve.

As a result of streamline contraction in the liquid, a pressure drop will occur over the restriction. In the cross-section in which the contraction is the strongest (vena contracta), the pressure reaches a minimum value. A partial pressure recovery occurs downstream in the pipeline. The final pressure drop, Δp, over the valve can be expressed in the density of the liquid (ρ), the pipe area (A), the pipe flow rate (Q) and a dimensionless coefficient of losses (ξ), which depends on the geometry of the restriction and, to a lesser extent, on the viscosity and the Reynolds number.

A.6.2 Valve characteristics

The discharge characteristic gives the correlation between the flow rate Q through the valve and the pressure difference Δp over the valve as function of the valve position. This correlation is expressed in a loss coefficient, ξ, or in a discharge coefficient, K_v or C_v. Note that these loss coefficients only apply in the main flow direction. In the case of reverse flow, the loss coefficients can differ strongly.

The coefficients are derived from the general equation for a Newtonian turbulent flow through a restriction in a pipeline under cavitation-free conditions:

$$\Delta h = \xi \frac{v^2}{2g} \quad \rightarrow \quad \Delta p = \rho g \Delta h = \frac{1}{2}\xi \rho v^2 = \frac{1}{2}\rho \xi \frac{Q^2}{\left(\frac{\pi D^2}{4}\right)^2} \tag{A.42}$$

in which:

Q = Flow rate	[m^3/s]
ξ = Loss coefficient	[-]
D = Valve diameter	[m]
ρ = Density of liquid	[kg/m^3]
Δp = Pressure difference	[N/m^2]

or, written differently, in which A_v is the discharge coefficient in [m^2]:

$$Q = \sqrt{\frac{2\left(\pi/4 \cdot D^2\right)^2}{\rho \xi}}\sqrt{\Delta p} = \frac{1}{\sqrt{\xi}}\frac{\pi D^2}{4}\sqrt{\frac{2\Delta p}{\rho}} = A_v\sqrt{\frac{\Delta p}{\rho}} \tag{A.43}$$

Here, A_v can be considered a valve constant. In general, such a relation will apply in case of turbulent flow through a valve. Every geometric configuration, thus every type of valve and every valve position, produces a different A_v value.

In practice, we use the following standardised discharge coefficient (K_v) that is derived from A_v. The discharge coefficient (K_v) is defined for water with a density of 1000 kg/m^3 between 5 and 30°C, in other words, K_v provides the water flow rate in m^3 that flows through the valve in one hour at a pressure difference of one bar. In addition to K_v, C_v is defined as discharge coefficient for American units. The discharge coefficient C_v is mostly used in the English-speaking world (England, North-America, etc.) and in the oil and gas industry (worldwide). The metric coefficient K_v is usually used in countries that use *Le*

Système international d'unités (English: SI system). Further considerations in this book use the K_v value.

$$Q = K_v \sqrt{\Delta p} = 0.865 \cdot C_v \cdot \sqrt{\Delta p} \tag{A.44}$$

in which:

Q = Flow rate [m³/h]

Δp = Net pressure difference over the valve [bar]

K_v = Discharge coefficient $\left[{m^3}/{h}\sqrt{bar} \right]$

For C_v-values the following units are used:

Q =Flow rate [US gallons/min]

Δp = Net pressure difference over the valve [psi]

C_v = Discharge coefficient $\left[US_{gal}\sqrt{psi} \right]$

The following correlation exists between ξ and K_v:

$$\xi = 1.6 \cdot 10^9 \frac{1}{K_v^2} \cdot D^4 \tag{A.45}$$

The coefficients are defined for the nominal valve diameter and the upstream and downstream pipeline, regardless of the actual inner valve diameter. This means that for a DN200 butterfly valve the nominal diameter is 200 mm. The K_v or C_v coefficients of a DN200 valve can be compared with each other regardless of the type of valve or the effective passage of the valve. Correction factors are applied for deviating connecting pipeline diameters. How to determine the discharge characteristics is established in standards.

The discharge coefficients (K_v, C_v) and the loss coefficient (ξ) are a function of the valve opening. The K_v of an entirely opened valve ($K_{v,100}$) is a suitable parameter for a valve and is a good basis for comparing different types of valves. It is also used for selecting the type of valve and valve diameter in the design. The dimensionless loss coefficient, ξ, is mostly useful when calculating hydraulic losses for pipeline systems as a result of all sorts of flow disturbances, such as bends, diameter transitions, T-pieces, valves, check valves, etc.

In many cases, the pipeline diameters at the upstream and downstream side differ from the nominal valve diameter. Then it is advised to include the piping geometry factor, F_p, according to IEC Standard 534-2-1 as follows:

$$K_v = \frac{Q}{F_p \cdot \sqrt{p_{up} - p_{down}}} = \frac{Q}{F_p \cdot \sqrt{\Delta p}} \tag{A.46}$$

$$F_p = \frac{1}{\sqrt{1 + \frac{(K_1 + K_2 + K_{B1} - K_{B2})}{0.00214} \cdot \left(\frac{K_v}{DN_{valve}^{\;2}} \right)^2}}$$

$$K_1 = 0.5 \cdot \left[1 - \left(\frac{DN_{valve}}{DN_p}\right)^2\right]^2$$

$$K_2 = 1.0 \cdot \left[1 - \left(\frac{DN_{valve}}{DN_p}\right)^2\right]^2$$

$$K_{B1} = 1 - \left(\frac{DN_{valve}}{DN_{p,red}}\right)^4$$

$$K_{B2} = 1 - \left(\frac{DN_{valve}}{DN_{p,exp}}\right)^4$$

with

K_v	… valve flow coefficient	[m³/h/bar]
Q	… flow rate	[m³/h]
p_{up}	… pressure upstream of control valve	[bar]
p_{down}	… pressure downstream of control valve (set pressure)	[bar]
Δp	… differential pressure across the control valve	[bar]
F_p	… piping geometry factor	[-]
K_1	… resistance coefficient due to head losses at inlet	[-]
K_2	… resistance coefficient due to head losses at outlet	[-]
K_{B1}	… Bernoulli coefficient due to reducer	[-]
K_{B2}	… Bernoulli coefficient due to expander	[-]
DN_{valve}	… nominal valve diameter	[mm]
DN_p	… nominal pipe diameter	[mm]
$DN_{p,red}$	… nominal pipe diameter of reducer	[mm]
$DN_{p,exp}$	… nominal pipe diameter of expander	[mm]

The piping geometry factor F_p represents an empirical correction factor for fittings (reducers, elbows, tees) directly attached to inlet and outlet connections of the control valve. In case the nominal control valve diameter has the same size as the pipe diameter (upstream and downstream), the valve fitting factor becomes $F_p = 1.0$.

A.6.3 Check valves

A check valve is a device that consists of a valve housing around one or several rotating or translational shut-off bodies with the purpose of allowing liquid to pass only in one direction (Figure A.33).

The liquid primarily controls the movement of the shut-off bodies. In some types, this movement is also influenced by a damping mechanism, which slows down the movement of the shut-off bodies (usually only during the last part of the shut-off phase).

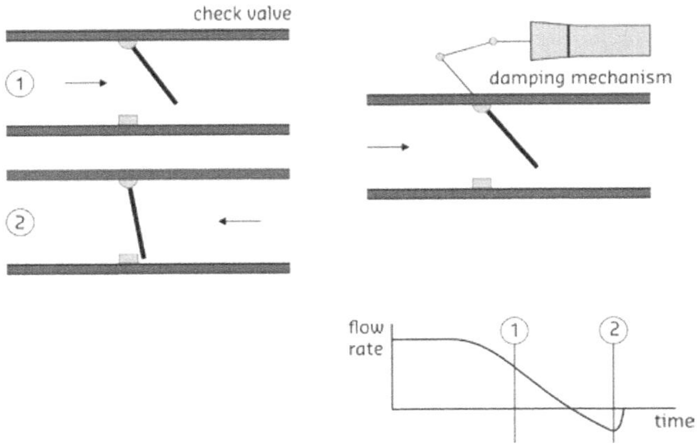

Figure A.33 Closing behaviour of check valves.

An ideal check valve will close in a decelerating flow at the moment that the flow rate has become zero. However, in practice, a check valve will close in a more or less developed return flow rate as a result of inertia and friction effects. Reduction of the return flow rate to zero (valve closed) goes hand-in-hand with water hammer phenomena. This leads to pressure increase on the (initial) downstream side and to a pressure decrease on the (initial) upstream side. The level of these pressure changes depends on the maximum return-flow velocity (v_R), the duration in which v_R, decreases to zero, and whether or not reflections of pressure surges occur during closing. These pressure changes create forces on the pipeline system and the anchoring/supports.

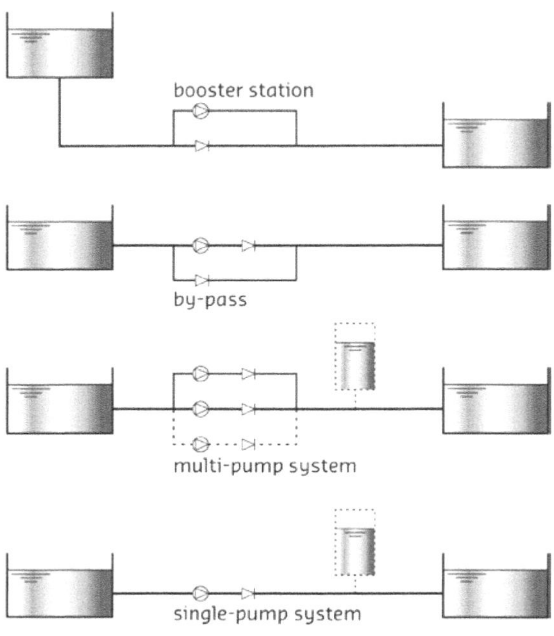

Figure A.34 Installation situations.

Common applications for check valves are (Figure A.34):

- Protecting the pump and motor against a too high return flow or reverse rotation in case of pump failure.
- Preventing return flow as a result of parallel pumps in the pump station that are not in operation.
- In a by-pass configuration, the check valve is used as a water hammer surge protection to suppress the creation of cavitation in the main.
- Preventing return flow in the main with a booster station that is in operation.
- Sealing a pipeline in case of pipe fracture to prevent the escape of liquid (for example cooling water circuits of nuclear power stations).

Check valves are divided into two categories:

- Undamped check valves: Reduction of the maximum return velocity to zero occurs briefly (Δt_R is in the order of 10^{-3} sec). The closing movement and the pressure changes are not influenced by reflections of pressure surges.
- Damped check valves: Reduction of the maximum return velocity to zero occurs rapidly for slightly damped valves (Δt_R is in the order of 10^{-3} up to 10^{-1} sec) and slowly for highly damped valves (Δt_R is in the order of 1 sec). The closing movement and the pressure changes can be influenced by reflections of pressure surges.

If the reduction time Δt_R is more than two seconds, one rarely needs to use a check valve, but rather a regulating valve or an operated valve.

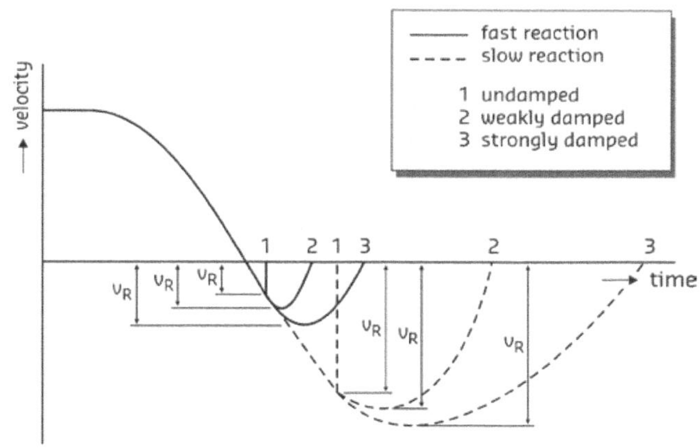

Figure A.35 Check valves classified according to their behaviour.

The dynamic behaviour of undamped and damped check valves can be characterised by:

- Rapidly reacting valves: the maximum return velocity, v_R, is relatively low.
- Slowly reacting valves: the maximum return velocity, v_R, is relatively high.

A.6.4 Characterising check valves

The stationary behaviour of check valves is characterised by the stationary characteristic in which the correlation between the pressure difference over the valve and the flow rate is given. The pressure difference over the valve is described in Equation (A.12) and repeated below:

$$\Delta H = \xi \frac{v^2}{2g}$$

The dimensionless resistance coefficient or loss coefficient, ξ, is a function of the valve position, θ. ξ is more or less constant for turbulent stationary currents with high Reynolds numbers, see Equation (A.6).

In the stationary characteristic, the correlation is given between pressure difference/pressure level difference and liquid velocity/flow rate. In Figure A.36, in which the pressure level difference is plotted as a function of the liquid velocity, the characteristics of a translational (nozzle) type of check valve are presented for two different springs. The pressure difference (Δh_{cl}) is the value at which the valve begins to open. The critical velocity (v_o) is the liquid velocity at which the valve is completely open. The characteristic is parabolic for liquid velocities larger than the critical velocity, because ξ is constant for a completely opened valve. When the flow rate decreases, the check valve begins to close from the moment that the critical velocity v_o is not exceeded anymore. Here, as a result of a resistance increase, the pressure difference/pressure level difference over the valve can increase slightly.

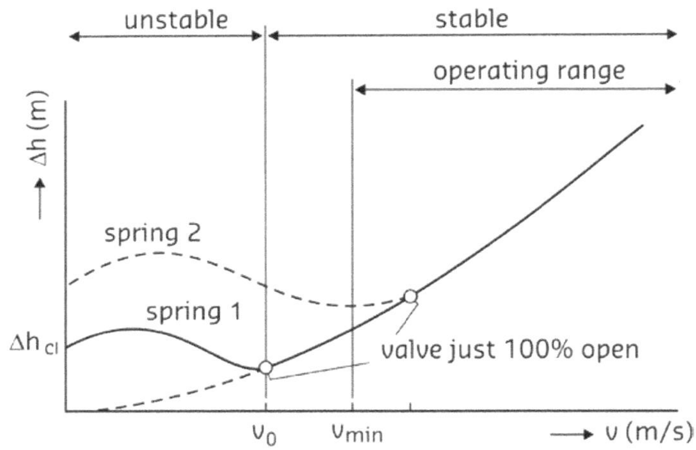

Figure A.36 Stationary characteristic.

Depending on the type of valve, the stationary characteristic may be influenced by springs (stronger, weaker), weights and balls (heavier, lighter), and the stroke or maximum rotation angle of the moving parts.

The closing behaviour of undamped check valves is described in the 'dynamic' characteristic:

$$v_R = f\left(\frac{dv}{dt}\right) \tag{A.47}$$

A characteristic liquid deceleration is plotted on the horizontal axis. The maximum return velocity is plotted on the vertical axis, which usually occurs immediately before the moving components reach the seating, see Figure A.37.

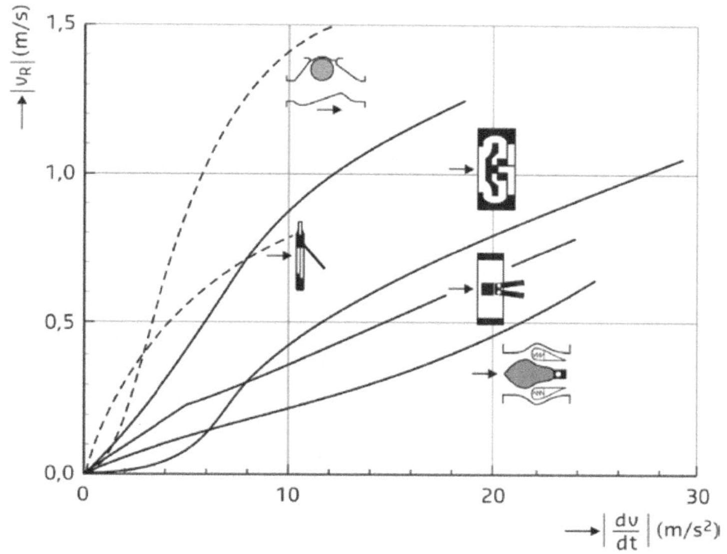

Figure A.37 Dynamic characteristics of DN200 check valves for horizontal installation.

An ideal check valve will close the moment that the flow rate has become zero, without the sealing causing extra pressure variations and without the occurrence of water hammer. This behaviour should be independent of the liquid deceleration. Therefore, the dynamic characteristic of an ideal check valve coincides with the horizontal axis of Figure A.35.

Depending on the type of valve, the dynamic characteristic is influenced by the valve diameter, springs (stronger, weaker), weights and balls (heavier, lighter), and the stroke or maximum opening angle of the moving parts.

The dynamic characteristic can also be used in a dimensionless form:

$$\frac{v_R}{v_o} = f\left(\frac{D}{v_o^2}\frac{dv}{dt}\right) \tag{A.48}$$

In which:

v_R = Maximum return velocity [m/s]

v_o = Critical velocity [m/s]

D = Internal diameter [m]

$\frac{dv}{dt}$ = Characteristic liquid deceleration [m/s^2]

Depending on the type of valve, the dimensionless dynamic characteristic may be influenced by weights and balls (heavier, lighter) and the stroke or maximum rotation angle of the moving parts, but not by the valve diameter and parameters that are only represented by the critical velocity (such as springs).

Figure A.38 shows measured dynamic characteristics of a translational (nozzle) type 800 mm check valve for three different springs. The same measurement data are shown in Figure A.39 in dimensionless form.

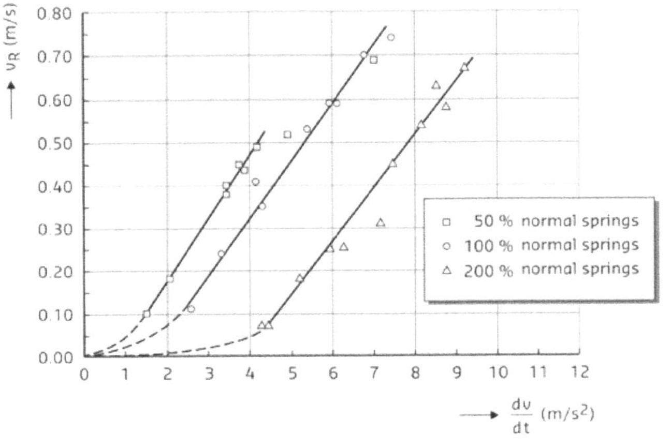

Figure A.38 Dynamic characteristic of a DN800 translational (nozzle-type) check valve for three different springs.

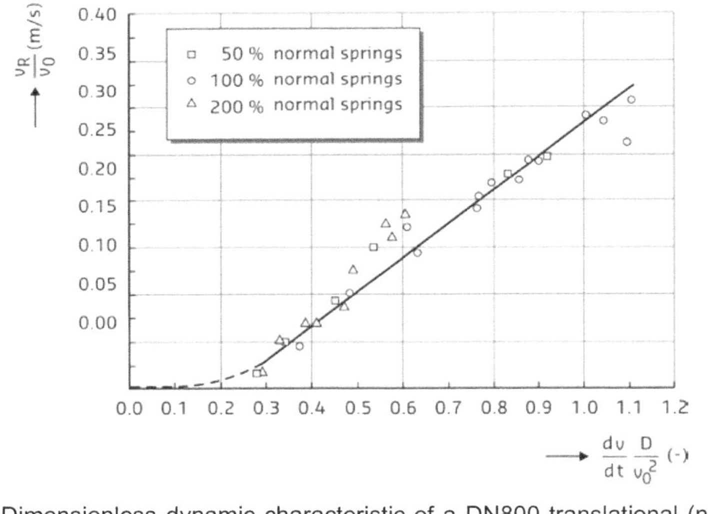

Figure A.39 Dimensionless dynamic characteristic of a DN800 translational (nozzle-type) check valve for three different springs.

A.7 WATER HAMMER IN PIPELINE SYSTEMS

Water hammer is a physical phenomenon that can occur in pipelines that are entirely filled with liquid. Changes in the flow rate always result in pressure changes that are transmitted through the pipeline as positive or negative pressure waves. These pressure surges are called water hammer. Both types of pressure surges can lead to a situation in which the pipeline strength becomes an issue. Negative pressure can also cause cavitation if the pressure in the liquid drops to vapour pressure.

Velocity changes can be the result of opening/closing of shut-off valves and starting/stopping of pumps. Water hammer can also occur as a result of venting a pipeline.

Water hammer can occur in every type of pipeline system and transport liquid: long and short systems with large and small diameters in single and branched networks for transporting oil, sewage water, drinking water, cooling water, dredge spoil, etc.

A.7.1 Propagation speed

The pressure wave speed can be derived from the continuity equation and represents all flexible storage in the pipeline, the compressibility of the liquid, and the radial and axial expansion of the pipeline:

$$\frac{1}{c^2}=\rho\left(\frac{1}{K}+\frac{1}{A}\frac{dA}{dP}+\frac{1}{\Delta s}\frac{d\Delta s}{dP}\right)=\rho\left(\frac{1}{K}+\frac{D}{eE}\right) \tag{A.49}$$

in which:

c = propagation speed of pressure wave	[m/s]
ρ = Density	[kg/m^3]
K = Compression modulus	[N/m^2]
A = Surface of the pipeline's cross section	[m^2]
P = Pressure	[Pa]
Δs = Pipeline element	[m]
D = Pipeline diameter	[m]
e = Wall thickness	[m]
E = Elasticity modulus (Young's modulus)	[N/m^2]

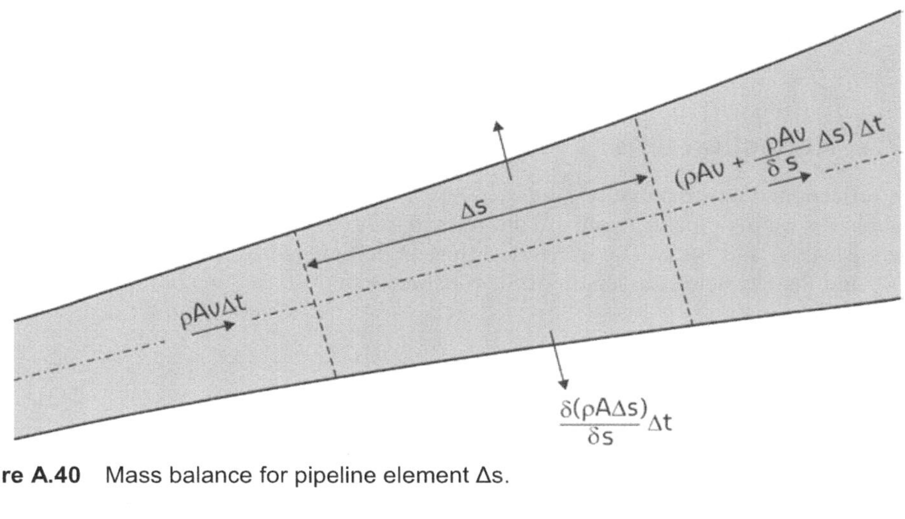

Figure A.40 Mass balance for pipeline element Δs.

The elastic properties of a pipeline can be summarised in the water hammer storage, λ_0 [m^2]:

$$\lambda_0=\frac{\Delta V}{\Delta H}=\frac{g\cdot V}{c^2} \tag{A.50}$$

In which:

ΔV = Volume change [m^3]
ΔH = Change in pressure level [m]
g = Gravitational acceleration [m/s^2]
V = Pipeline volume [m^3]
c = propagation speed of the pressure surges [m/s]

The aforementioned equation only applies if there is no air in the pipeline. If there are gas bubbles in the pipeline, a term should be added to Equation (A.49). The water hammer storage in the pipeline changes, because the gas bubble compresses and expands under the influence of the pressure according to the ideal gas law:

$$P \cdot V^k = C \tag{A.51}$$

in which:

k = Polytropic constant [-]

or

$$A_{gas} = \frac{\Delta V}{\Delta H} = \frac{\rho \cdot g \cdot V_i}{k \cdot P_i} \tag{A.52}$$

The total water hammer storage in a system with gas bubbles is the sum of Equations (A.50) and (A.52). From this, an average propagation speed can be derived:

$$\lambda_0 + A_{gas} = \frac{\Delta V}{\Delta H} \equiv \frac{g \cdot V}{c^2} + \frac{\rho \cdot g \cdot V_i}{k \cdot P_i}$$

$$\frac{1}{c_{gas}^2} = \frac{1}{c_0^2} + \frac{\rho \cdot V_i}{k \cdot P_i \cdot V} \tag{A.53}$$

A.7.2 Reflection time

The reflection time of pressure surges, μ [s], is the time a pressure surge takes to run through the entire pipeline/main of the system from the initiator of the pressure surge (closing valve) and back. This reflection time is determined by the propagation speed, c [m/s], and the characteristic length of the pipeline system, L [m], according to the formula (see also Figure A.41):

$$\mu = \frac{2L}{c} \tag{A.54}$$

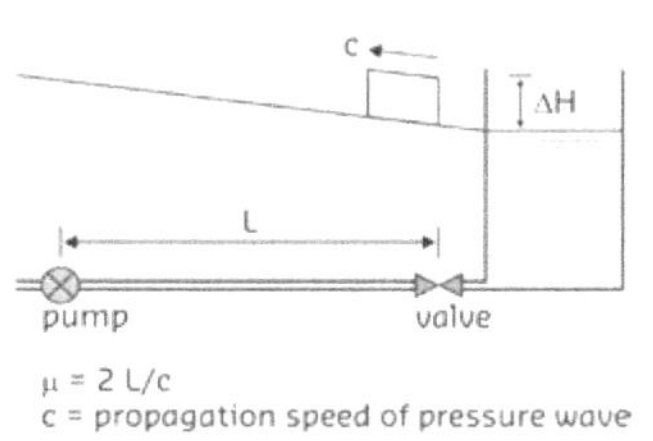

Figure A.41 The concept 'propagation time'.

We refer to quick phenomena if the disturbance (e.g. valve closure) occurs within the reflection time of the pressure surge. We refer to slow phenomena if the disturbance is 5–10 times slower than the reflection time.

A.7.3 Joukowsky's Law

The greater and faster the velocity changes, the greater the pressure changes. Here, 'fast' is related to the reflection time of pressure surges μ. Joukowsky's Law states the relationship between the magnitude of the velocity change (Δv [m/s]) and the magnitude of the change in pressure level (ΔH [m]):

$$\Delta H = \pm \frac{c}{g} \Delta v \tag{A.55}$$

Figure A.42 explains this formula.

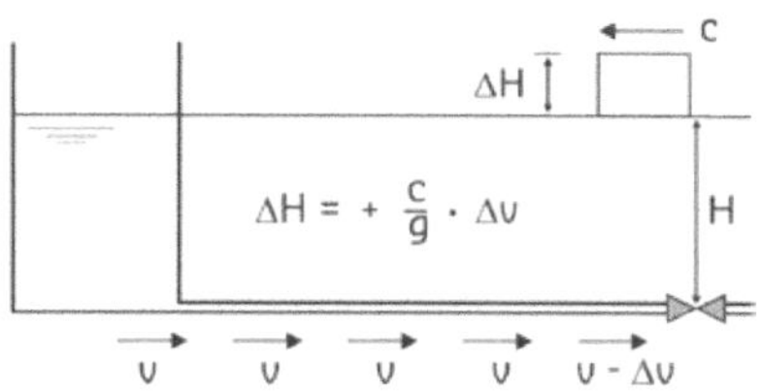

Figure A.42 Joukowsky's Law.

$\Delta H \approx 100\ \Delta v$ applies to steel and concrete pipelines ($c \approx 1000$ m/s) filled with water. This means that a pressure change of 10 metres water column (~1 bar) will occur at a relatively small velocity change of 0.1 m/s.

This relationship only applies if the velocity change occurs within the reflection time and the friction loss is small with respect to the pressure in the pipeline. However, this is not always the case in practise. Furthermore, this rule of thumb has limited validity in branched systems and diameter transitions due to local reflections.

Water hammer phenomena can be calculated precisely using the equation of motion, the law of conservation of mass (= principle of mass conservation) and the related preconditions (pipeline details, pump data, levels in the reservoirs, etc.).

A.7.4 Allowed pressure in synthetic pipelines

The maximum allowed positive and/or negative pressure in a pipeline depends on the type of material. This section is confined to the most commonly used materials: polyvinylchloride (PVC) and high-density polyethylene (HDPE).

Barlow's equation provides the maximum allowed positive pressure P_{max} in a pipeline (Wijdieks 1978):

$$P_{max} = \frac{2\sigma}{\frac{D_e}{e} - 2} \tag{A.56}$$

In which D_e is the outer diameter and σ is the allowed tangential wall tensile stress. The maximum allowed tangential tensile stress depends on the pipeline material.

The following equation provides the maximum allowed negative pressure P_{min} (formally, the net external pressure expressed in *bar gauge*) in a pipeline:

$$P_{min} = -2E\left(\frac{e}{D_i}\right)^3 \tag{A.57}$$

In which D_i is the inner diameter and E is the elasticity modulus (Young's modulus) of the material. Exceeding this maximum allowed negative pressure deforms the material.

Table A.7 Material characteristics of synthetic pipelines.

Material	Elastic Modulus (10^9 N/m^2)	Maximum Tangential Tensile Stress (10^6 N/m^2)
PVC	3.0	12.5
HDPE (different types)	1.0	5.0–8.0

A.7.5 Surge vessel as water hammer surge protection

The surge vessel is a pressure vessel that is partially filled with gas (air) and water. The gas volume is in fact a compressed spring that pushes the water into the pipeline after a pump failure. It is as if the surge vessel takes over the function of the pump (delivery of water), which reduces the flow rate more slowly and thereby, leading to a reduction of the dynamic pressure.

The gas volume and the gas pressure determine the behaviour of the surge vessel and therefore, they should be maintained. To guarantee the gas volume and pressure, the surge vessel is constructed as a vented vessel or as a non-vented vessel including a compressor. If the surge vessel is constructed as a vented vessel, the air valve must be above the supply pressure level so that the vessel is always vented in case of pump standstill: the gas volume is now refreshed in each pump cycle. If the surge vessel is constructed as a non-vented vessel, a compressor should control the gas volume.

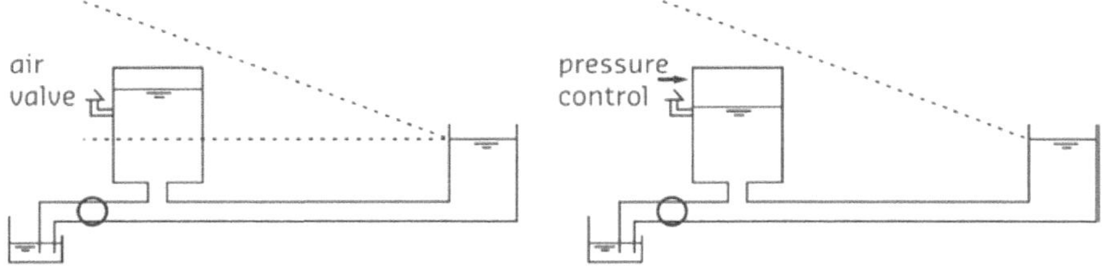

Figure A.43 Vented and non-vented vertical surge vessels.

In addition, one has the option to select a horizontally or vertically installed surge vessel. This choice depends primarily on constructional and technical considerations (location, available height, etc.). The difference in hydrodynamic behaviour is portrayed using computer simulations.

The most efficient operation of the vessel is achieved, if it is installed directly in the pipeline when flow passes through the air vessel continuously. The advantage of this is that the vessel can deliver immediately. It is also an ideal venting arrangement for air taken in by the pump. The disadvantage of a flow-through vessel is that debris can collect at the bottom and that the surge vessel cannot be taken out of operation easily for maintenance activities.

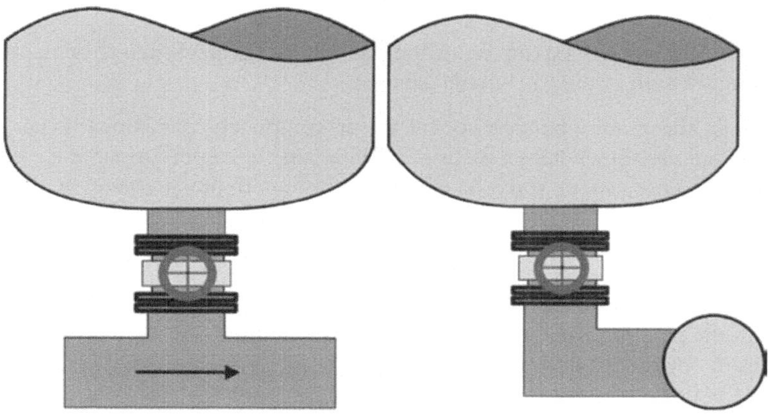

Figure A.44 Standard arrangements of a vertically installed surge vessel.

A better solution is to install the vessel on top of the pipeline and to connect it using a T-connection (with a diameter that is preferably equal to the diameter of the main).

Due to lack of space, it might be necessary to install the vessel alongside the pipeline and to connect it to the pressure pipeline using an internal suction pipeline. This suction pipeline should be kept as short as possible and have a diameter that is preferably equal to the diameter of the main pipeline. The efficiency decreases as the feed-through pipeline becomes longer and the resistance between vessel and main pipeline becomes bigger.

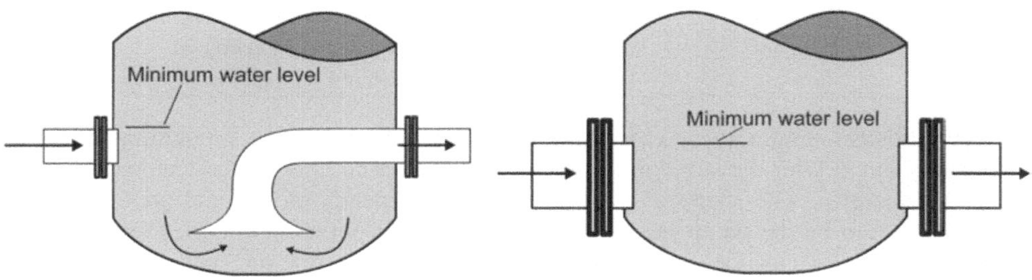

Figure A.45 Several layouts for a surge vessel.

The operation principle of the surge vessel for safeguarding a pipeline against negative pressure as a result of pump failure is as follows: The surge vessel is located near the pumping station. After a pump failure, the flow rate of the pump drops very quickly to zero. The vessel will deliver immediately. The pressure in the vessel is initially equal to the outlet pressure of the pump. Due to the delivery, the level in the vessel falls and, as a result, the pressure in the surge vessel drops and the gas volumes becomes bigger.

The surge vessel should be dimensioned in such a way that gas cannot penetrate into the main pipeline. Equation (A.50) determines the behaviour of the gas volume.

The polytropic (or Laplace) coefficient is a characteristic of the gas and determines the expansion behaviour of the gas bubble: $k = 1$ in case of isothermal behaviour and $k = 1.4$ in case of adiabatic behaviour.

- In case of isothermal expansion, heat exchange to the surroundings occurs so that the gas temperature does not rise.
- In case of adiabatic expansion of the gas bubble, heat exchange does not occur and the temperature of the gas bubble can rise.

In practise, the mode changes do not occur completely adiabatically nor completely isothermally but are somewhere in between. These are called polytropic mode changes.

The vented surge vessel starts to allow air inflow if the pressure in the gas bubble becomes lower than the atmospheric pressure. Subsequently, the surge vessel will behave as a surge tower.

The expansion behaviour influences the results. In case of isothermal behaviour, the lowest level in the surge vessel is reached. In case of adiabatic behaviour, the lowest pressures in the pipeline are achieved. Therefore, both situations are calculated in hydraulic simulations to obtain a conservative design.

Installing a surge vessel influences the dynamic behaviour of the check valve. In case of pump failure, the delay of the water column between the suction side and the surge vessel increases tremendously, because the pressure behind the valve is 'held' by the surge vessel. This means that a fast-closing undamped check valve or a damped check valve should be used.

In general, dimensioning a surge vessel requires several calculations because there are several combinations to cover the entire dynamic behaviour. The following table provides an overview of the calculation possibilities for demonstrating and dimensioning a surge vessel in case of pump failure.

Resistance of Pipeline	**Suction and Press level**	**Behaviour of Surge Vessel**	**Result**
High	Low	None	Minimum pressure in pipeline
High	High	None	Maximum pressure in pipeline
Low	Low	Isothermal	Minimum level in vessel
High	Low	Adiabatic	Minimum pressure in pipeline

When dimensioning a surge vessel, the duration of the hydraulic simulation should be sufficiently long. After the subsequent delivery, the water column will swing back causing the vessel to refill. The speed at which this occurs depends on the end-pressure in the system. This can be the pump pressure of a connecting pipeline in complex systems. The gas volume should cushion this oscillation with which the pressure can again increase considerably. Therefore, the duration of the simulation must cover the complete delivery cycle of the surge vessel with a minimum and maximum level in the surge vessel.

A.7.6 Surge tower as water hammer surge protection

A surge tower is simply an open tank with sufficient surface area that slows down quick fluctuation in flow rate occurring in the main pipeline, for example due to pump trip. If the surge tower is placed at the upstream part of the pipeline, it supplies water to the pipeline in case of pump failure and the level in the tower falls gradually. The surge tower, which is placed at the downstream part of the pipeline, absorbs water when a valve is closed and the level in the tower rises.

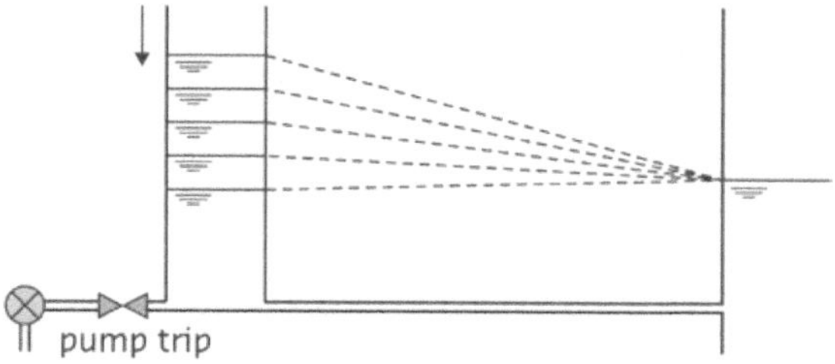

Figure A.46 Effect of a surge tower during pump failure.

Initially, the flow rate takeover by the surge tower causes the flow rate of the pipeline to remain stable. Subsequently, the velocity in the pipeline drops gradually to zero due to the fall or rise of the tower level.

A simple water tower has only a small inflow and outflow resistance. In general, it will be dimensioned in such a way that it will not run empty, in case of maximum required flow. If a smaller flow is consumed downstream, it will not overflow. However, an overflow chamber is installed in some cases to allow overflowing.

The big **advantage** of the surge tower is its simplicity. Because it lacks mechanical elements, it is a reliable tool. The level in the tower adjusts itself automatically to the locally prevailing pressure. The buffer tower does not require monitoring and has low maintenance.

Disadvantages of the surge tower are that the usage is limited, because the height of the tower must be equal to the pump head ($Q = 0$). There are also esthetical objections to such a high construction. In case of wastewater pipelines, using a surge tower can result in stench nuisance and, if the capacity of the pipeline system is expanded in the future, the capacity of the buffer tower cannot be enlarged easily.

A.7.7 Bypass as water hammer surge protection

A bypass pipeline can be used to limit negative pressures in case of pump failure. In order to let the by-pass pipeline make a quick subsequent delivery, a reservoir should be available at a sufficiently high level.

A by-pass pipeline that uses the same suction basin as the pump has the **disadvantage** that it only makes a subsequent delivery after the pressure downstream of the pump has dropped below the suction level. The use of a by-pass pipeline is also limited to pipelines of smaller lengths. **Advantage** is that special requirements are not set to the check valve behind the pump.

A.7.8 Air valve as water hammer surge protection

Air valves can be used to limit unacceptable negative pressures in the pipeline system after pump failure. As soon as the pressure at the location of the air valve drops below the atmospheric pressure, the air valve opens and creates air inflow. The result is that the local pressure no longer drops significantly.

Following the opening of the air valve, it is important that the entrained air can be removed from the system. For this purpose, air valves are often fitted with an air release mechanism. The air inlet capacity is usually many times larger than the air outlet capacity.

This is to release air slowly in order to prevent a high pressure surge at the end of the venting process.

The use of air valves as water hammer surge protection contradicts the design advice to prevent air entrainment into the system. Therefore, using air valves in situations in which negative pressures occurs during regular operations must be discouraged. However, in incidental cases, it should be considered an option for protecting the pipeline.

A number of remarks are given:

- The entrained air can be transported past a low point, which means that the air will no longer escape at the same location as it was entrained. Therefore, an additional air valve further downstream might be necessary. Otherwise, this leads immediately to extra energy loss.
- In case of pump standstill, the system must be pressurized at the location of the air valve to enable the air release process.
- The air valve should be well dimensioned, see recommendations in Section 4.6 in the main text.
- Depending on the construction of the air valve, this equipment is sensitive to clogging if it is only operated incidentally. Floating dirt and grease can clog the mechanism, which causes it to malfunction. Therefore, solid designs and/or frequent maintenance are required.
- Air release causes stench nuisance.

A.8 GAS BUBBLES IN PIPELINES

A minimum flow velocity is required to start a gas transport in a declining pipeline under an angle θ. If this velocity is lower than the minimum velocity, the air in the pipeline will accumulate until the bubble reaches the bottom of the declining pipeline. In that case, the maximum additional energy loss should be taken into account. This maximum additional energy loss is equal to the difference in height over the declining pipeline.

However, the necessary flow velocity also depends highly on the pipeline diameter. That is why we use a dimensionless flow velocity – the flow number:

$$F = \frac{v}{\sqrt{gD}} \tag{A.58}$$

The flow number allows for phenomena that are dominated by gravity to be transferred to other pipeline diameters. The free surface flow under a gas bubble with a hydraulic jump is an example of this. The flow number does not take into account the transfer effects with regard to the water quality, but research has shown that these transfer effects are negligible in pipeline diameters larger than 150 mm (Tukker, 2007).

Practical recommendations regarding air in wastewater pressure pipelines, derived from CAPWAT research, are summarised in several publications (Pothof *et al.* 2011a, Pothof *et al.* 2011b).

A.8.1 Transport mechanisms

The elementary forces for air bubble transport in water in declining pipeline sections are as follows:

- Buoyant force
- Resistance
- Balance between surface tensions (water / air bubble / wall)

Many researchers have focussed on studying the details. The CAPWAT project has looked into the phenomenon on a large scale. Lubbers (2007) has researched for the first time the influence of the pipeline diameter and the length of the declining section on air transport through streaming water. Pothof (2011) completed the CAPWAT project and presented a comprehensive study on the capacity reducing effect of air pockets. In addition, Pothof et al. (2013) published research on the effect of surface tension on air water flows.

To this end, several test circuits (mainly made of Perspex) were built with varying diameters (∅80 mm – ∅500 mm) and varying slope angles (5°C – 90°C). At the upstream side of the pipeline, air was injected and monitored. Subsequently, it was observed how the injected small air bubbles behave under different flow velocities. This process is described in the following in detail using illustrations and photographs.

Small air bubbles are released from the injection location. Water transports these air bubbles up to the bend. If the speed of the water is sufficiently high, the air bubbles pass the bend freely and flow through the sloping section to the deeper, horizontal section. Subsequently, they can continue to flow through the downstream ascending pipeline section of the pipe.

The speed of the air bubbles in the declining section is lower than the velocity of the water. In case of a critical water velocity and a 10° angle of slope, the air bubble speed amounts to approx. 60% of the water velocity for small air bubbles up to 30% of the water velocity for larger gas volumes.

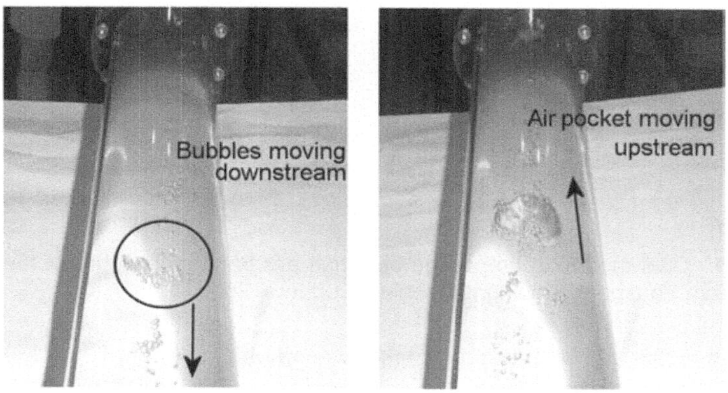

Figure A.47 Small bubbles flow downstream (left) and large gas volumes flow against the stream (right).

If the flow velocity is reduced, the air bubbles in the sloping section can accumulate into larger air bubbles that flow upstream (if they are large enough) due to the increased ratio between the buoyant force and the resistance. This air volume flows upstream to the bend and creates a 'buffer'. The result is that smaller air bubbles, transported from the horizontal section, can no longer flow freely through the bend, but get stuck in the 'buffer'. This causes stagnation of the air bubble transport and subsequently, the air bubble in the bend will become larger and larger. In Figure A.48, the left photo shows different gas bubbles that are transported together with the flow through the bend. The bubbles do not merge and the flow exerts sufficient force on the bubbles to transport them through the declining pipeline. The right photo shows a gas bubble that stands still at the high point. The flow speed is not able to exert sufficient force on the gas volume to remove it in one go. The gas volume functions as a buffer, blocking the transport of gas to the downstream section.

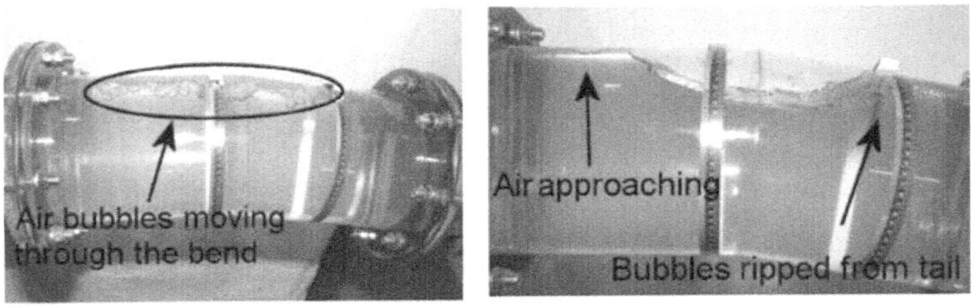

Figure A.48 Left: air bubbles moving through bend. Right: Stagnant gas bubble. (Lubbers, 2007).

Turbulence at the toe of the buffer causes the release of small bubbles that are subsequently transported in downstream direction. The level of turbulence is determined by the size of the bubble and the local velocity. The transported small air bubbles can merge once more into a large bubble and return to the bend against the stream, if the buoyant force is stronger than the forces that the flow exerts on the bubble. This results in a chaotic process in which a stream transports large air volumes/plugs upstream (against the flow direction), while a second stream transports smaller bubbles downstream.

Figure A.49 Dual-mode stream: Large air plugs are transported against the stream, while another stream transports small gas bubbles (Lubbers, 2007).

The result is that the net transport might be negligible, causing the air bubble to expand in the bend.

A.8.2 Velocity criterion

In case of a low water velocity, the air bubble will stay in the horizontal section. A flow regime is generated that is comparable with free outflow in an open-channel flow. The layer thickness, y_c, near the bend cannot become smaller than one in critical outflow (Froude number = 1) and, therefore:

$$\frac{v}{\sqrt{g A_{w,c}/b_{s,c}}} \approx \frac{v}{\sqrt{g y_c}} = 1 \qquad \text{(A.59)}$$

in which:

$A_{w,c}$ = Wet cross-section at critical depth of the water [m^2]

$b_{s,c}$ = Width at water surface at critical depth of the water [m]

At the upstream side, the air bubble cannot continue to expand infinitely. Friction and acceleration of the water under the air bubble determine the course of the water plane. The maximum bubble length $L_{max,hor}$ in the upstream horizontal section can be estimated from the hydraulic gradient in the entirely filled pipeline (valid for flow numbers that are smaller than 0.58).

$$L_{max,hor} \leq \frac{D - y_c}{\sin\alpha + HGL} \quad (A.60)$$

In which:

α = slope angle of the upstream pipeline ($\alpha > 0$ is an ascending pipeline) [°]

The bubble length (up to a maximum length of 7D) is significantly smaller for higher flow numbers ($F > 0.58$), see Hager (1999).

In case of an increase in velocity, the air bubble is forced to enter the declining section. The result of this is that the height of the gas volume in the sloping section increases and this creates additional energy loss.

The flow in the sloping section is supercritical. At the toe of the gas volume, a turbulent hydraulic jump is created in which large quantities of small air bubbles are taken in and transported onward. Depending on the flow velocity, three different flow images can occur (see Pothof & Clemens (2011) and Figure A.50).

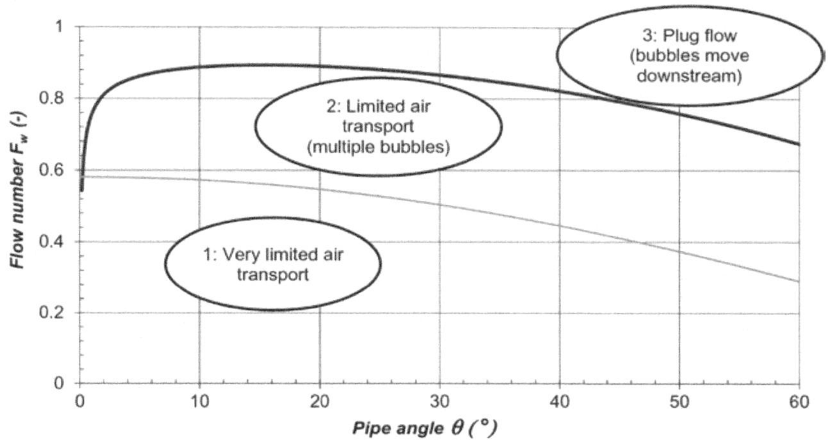

Figure A.50 Required flow numbers for gas discharge.

1. **Blow-back flow-regime[1]:** The air bubbles merge quickly at the top of the pipeline and the resulting gas volumes blow back upwards into the larger air volume. The declining pipeline contains exactly one long gas volumes with hydraulic jump.
2. **Transitional flow regime:** The air bubbles merge but they are not entirely returned in one large air bubble. Instead of this, a next gas volume is created, it remains

[1]The mentioned limiting value applies to pipeline diameters larger than 200mm and slightly lower limiting values apply to smaller diameters.

hanging in the declining pipeline and a next hydraulic jump develops. This process can repeat itself up to 7 times over 40 metres.

3. **Plug flow regime[2]:** A hydraulic jump is no longer in the declining pipeline, but all entrained air is immediately transported as small bubbles along the top of the pipeline. If the flow velocity is increased further, increasingly more bubbles are transported as suspended bubbles. This flow regime is called a 'dispersed bubble flow', but this flow regime will hardly occur in wastewater pressure pipelines because of the need of high flow speeds.

In principle, the bubble can fill the entire sloping section if there is sufficient air transport and therefore, the total energy loss is equal to the depth (height level difference) of the crossing pipe.

In case of increasing the water flow rate and therefore the velocity, the air bubble transport through the sloping section increases and as this is more than the supply, this means a reduction of the air bubble size.

A.8.3 Calculation model for gas transport

The calculation model described here is based on the dissertation of Pothof (2011). The free surface flow at normal depth in the sloping pipeline plays an important role in the construction of the mathematical model. The dimensionless expression for the free surface flow at normal depth in a round pipe is:

$$F_w^2 = \frac{v_{s,w}^2}{gD} = \frac{2\sin\theta}{\lambda}\frac{D_h}{D}\left(\frac{A_{w,n}}{A_D}\right)^2 \tag{A.61}$$

in which:

$A_{w,n}$ = Wet cross-section at normal depth [m]

D_h = Hydraulic diameter of the water film at normal depth [m]

At the normal depth, friction is in balance with gravity. The friction factor is calculated using White-Colebrook, but you can also use another friction formula.

The measuring data of the laboratory shows that a length of 9D is needed to reach the normal depth. In practise, each air accumulation is longer than 9D. Therefore, the change in length of the air bubbles can be based on the axiom that the water film always has a normal depth. Using this axiom, we find the relationship between the airflow rates and the change in length of the air bubbles:

$$Q_{g,in} - Q_{g,t} = \left(A_D - A_{w,n}\right)\frac{dL_g}{dt} \tag{A.62}$$

in which:

$Q_{g,in}$ = Influent air flow rate at the bottom of the declining pipeline [m^3/s]

$Q_{g,t}$ = Transported air flow rate at the bottom of the declining pipeline [m^3/s]

[2]The transition to full gas flow transport depends on the piped-in quantity of gas. The dimensionless flow rates of gas are $F_g = 0.3 - 0.0045$.

Equation (A.62) can also be used as basis for calculating the duration to degrade a gas bubble of a certain length. For this purpose, Equation (A.62) should be integrated over time:

$$\int Q_t - Q_{in} = (A_D - A_n)\cdot L_g \tag{A.63}$$

The shortest degrading time is found if we assume that the incoming air flow rate is $Q_{g,in}$ = 0 l/min. Equation (A.62) is made dimensionless.

$$F_{g,in} - F_g = \frac{(1 - A_{w,n}/A_D)}{\sqrt{gD}}\frac{dL_g}{dt} \tag{A.64}$$

The remainder of the description in this section consists of modelling the air flow number, F_g. Initially, the model is validated using measuring data at Eötvös numbers above 5000 (i.e., Hook of Holland data of the laboratory for $D > 200$ mm) and angles of slope $\theta \leq 30°$. For further explanation on the influence of the Eötvös number, the reader is referred to Pothof (2011):

In the following, the air transport in the plug flow regime with random slope angles is described. Subsequently, the impact of the air accumulation is taken into account.

Air transport in the plug-flow regime

Pothof (2010) had set up a balance of forces that predicts when a long gas bubble in a declining pipeline remains hanging under the influence of slip force and buoyant force, under the assumption that Eo > 5000.

$$\begin{aligned} F^2(\theta) &= \frac{\sin\theta}{\lambda}\frac{D_h}{D}\left(\frac{A_n}{A_D}\right)^2 \\ &+ \frac{A_D}{A_b}\frac{\cos\theta}{2\pi}\left[\frac{2}{3}\sqrt{\frac{2y_n}{R} - \left(\frac{y_n}{R}\right)^2}\left(\frac{y_n}{R} - 3\right)\left(\frac{y_n}{R} - \frac{1}{2}\right) + \arcsin\left(1 - \frac{y_n}{R}\right) + \frac{\pi}{2}\right] \end{aligned} \tag{A.65}$$

If the water flow number exceeds this value, then a single long air bubble will be transported downwards at a speed that can be determined with one of the drift flux models. Therefore, the equation (A.65) marks the transition to prop flow, except for a factor that depends on the airflow rate. The usefulness of the equation (A.65) in the blowback flow regime is illustrated in Figure A.51. If the water flow number is scaled once more with the critical flow number for individual air bubble transport according to Equation (A.65), then most of the results in Figure A.51 fall onto one line. Only the measured data in the smallest air flow rates (Fg-·1000 = 0.4) clearly deviate from this line, which is ascribed to the influence of aeration. We can conclude that the influence of the angle of slope can be modelled with equation (A.65).

Measurements show that the air transport of the plug flow regime increases exponentially as a function of the water flow number. The exponential trend can be best described as a function of the rescaled water flow number $F_w/F(\theta)$ in which $R^2 = 0.93$.

$$F_g = 1.87 \cdot 10^{-7} \cdot e^{9 \cdot F_w/F(\theta)} \tag{A.66}$$

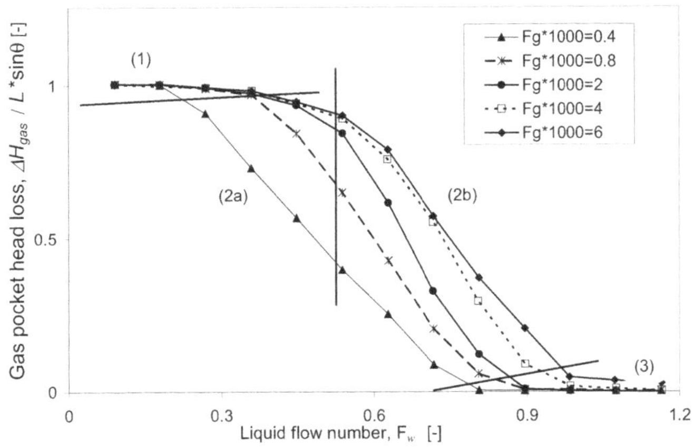

Figure A.51 Dimensionless air-bubble energy loss in Hook van Holland.

Equation (A.66) can be inverted in order to obtain an expression for the critical flow number of air bubble transport as a function of the airflow rate, diameter and slope angle.

$$F_c = F(\theta)\ln\left[\left(\frac{F_g \cdot 10^7}{1.87}\right)^{1/9}\right] \tag{A.67}$$

Equation (A.67) is valid for water-air flow in pipelines with $D > 190$ mm (or Eo > 5000). Equation (A.67) provides a more complete description than the current expressions for critical speed, because it also included the influence of the airflow rate.

Air transport model in the blowback flow regime

If there is accumulation of air in a declining pipeline, then the energy loss in equilibrium is described accurately based on a cumulative function of the beta-distribution $B(x,\alpha,\beta)$ in which x is the rescaled flow number and α and β are functions of the pipeline length.

$$1 - R = B(x,\alpha,\beta) = \frac{\int_0^x t^{\alpha-1}(1-t)^{\beta-1}\,dt}{\int_0^1 t^{\alpha-1}(1-t)^{\beta-1}\,dt} \tag{A.68}$$

in which

$$R = \Delta H_{gas}/(L\sin\theta)$$
$$x = F_w/F(\theta)$$
$$\alpha = 9.67 \cdot 10^{-2}(L/D - 10.3)^{0.783}, \quad 20 \le L/D \le 210$$
$$\beta = 9.39 \cdot 10^{-3}\, L/D + 0.439$$

The model in Equation (A.68) suggests that the inverse beta function in Equation (A.66) can be included in order to model the effect of the air accumulation.

$$F_g = 1.87 \cdot 10^{-7} \cdot e^{9 \cdot F_w / \left(B^{-1}(1-R,\alpha,\beta) \cdot F(\theta)\right)} \tag{A.69}$$

Equation (A.69) predicts the dimensionless gas flow rate as a function of the relevant parameters:

These model calculations are calibrated to the following circumstances:

- Pipeline lengths $L/D > 30$, which are the most relevant in practice. With shorter pipeline lengths, the assumption that the water film will reach the normal depth immediately is a far too large simplification. The maximum experimental pipeline length is L = 209*D. The theoretical criterion for gas bubble transport suggests that the air transport of declining pipelines will not decrease further. The model parameters α and β can be based on maximum $L/D = 210$; therefore $L/D = \min(L/D, 210)$.
- Pipeline diameter D > 190 mm. For smaller pipeline diameters, the actual dimensionless airflow rate is greater and the current model provides a conservative approach.
- If the calculated airflow rate becomes less than $F_g = 0.0005$, then the aeration of the water provides a significant contribution. This effect is not taken into account in this model, which means that the actual air flow rate will be somewhat greater.
- Most measurements are taken at a pressure of approximately 1.6 bar (absolute) at the bottom of the declining pipeline. If the system pressure is significantly higher, then the aeration plays an increasingly larger role. Several measurements at pressures of up to 2.5 bar (abs) show this effect. This effect is not included yet in the current model.
- During the measurements, the water temperature varied between 15°C and 25°C. In case of significantly lower temperatures of the wastewater, the airflow rate will drop further.

In spite of the aforementioned limitations, this calculation model provides a reliable indication of the air transport in declining pipelines. By substituting Equation (A.69) for Equation (A.64), we can calculate the decay speed of an air bubble. Subsequently, integration provided the necessary decay time for reducing an air bubble of a certain length to an acceptable length. Based on the decay time and the water flow rate, we can calculate the water volume necessary for decay of the air bubble. Figure A.52 shows that the calculation model provides an excellent indication of the breakdown process.

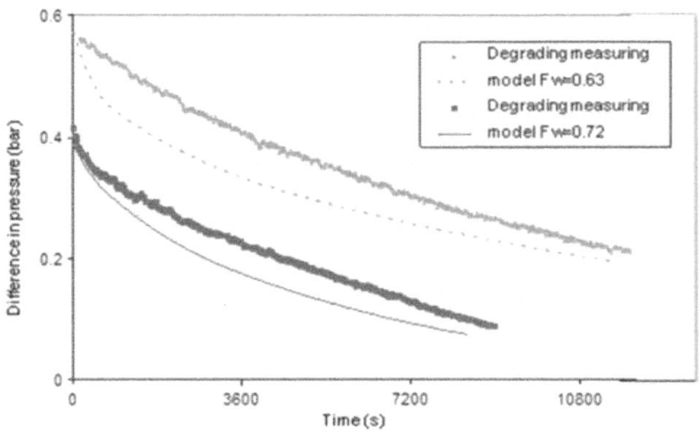

Figure A.52 Comparison between calculated and measured decay. The decay measurements are carried out at 11°C, which most likely explains the great differences in decay speed at the beginning of the measurement. Nevertheless, the calculation model gives a good indication of the decay process.

The calculation model shows that the airflow rate for longer declining pipelines ($L/D > 200$) depends exclusively on the relative bubble length L_b / L. Moreover, the airflow rate for declining angles between 5° and 20° is nearly constant. That is why the decay time for a certain declining pipeline length can be expressed as the function of gas bubble accumulation. Figure A.53 shows the decay time per 40m of declining length for a pipeline diameter of 192 mm at various water flow rates. This decay time can be scaled to larger pipeline diameters and longer declining pipelines. The decay time increases linearly to the pipeline length: a declining pipeline, which is 5 times as long, has a decay time that is 5 times as long. The decay time decreases in case of larger pipeline diameters by the square root of the diameter ratio; a pipeline diameter, which is 4 times as large (D=768mm), results in cutting by half the decay time per 40 m of pipeline length and a constant flow number. Based on these correction factors for the pipeline diameter and pipeline length, Figure A.53 can be used to indicate the expected decay time and the water volume required for this action.

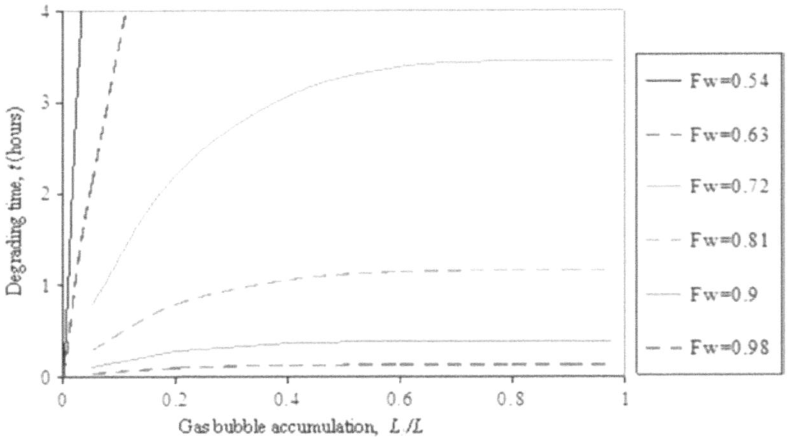

Figure A.53 Decay time as a function of the relative bubble length per 40m when D = 192mm and θ = 10–20°.

A.8.4 Catching gas

T-pieces can be used for catching gas bubbles that are carried to the top part of the pipeline by the flow. Gas bubbles that are carried to the middle or lower side of the pipeline by a turbulent flow cannot be caught with conventional air valves.

From research by Wickenhäuser (2008), who used clean water for applications in hydroelectric power stations, it emerges that the minimum dimensions of a T-piece are as follows (Figure A.54):

Width: 0.5 D

Height: 0.3 D

Wickenhäuser used a 500mm pipeline with an orthogonal construction for catching gas. By using an orthogonal form, more available surface was created on both sides for catching gas. The aforementioned numbers are already corrected for this difference.

Wickenhäuser's study was carried out in a ∅500 mm pipeline with a flow number of F < 0.62. This is equal to a flow speed of nearly 1.4 m/s.

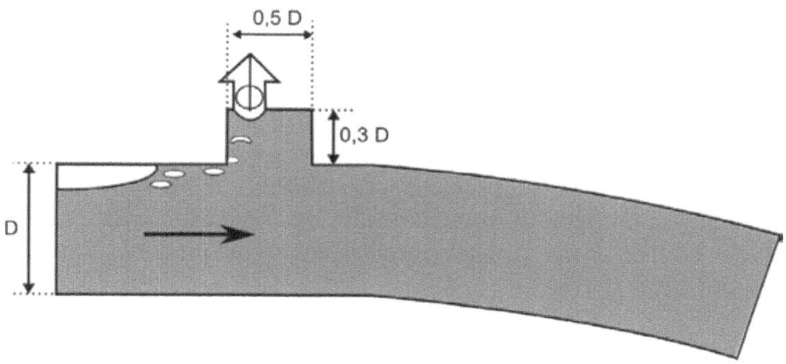

Figure A.54 Pipeline with a T-piece and air valve.

Figure A.55 (left) shows the flow pattern in a T-piece with H > 0.3 D. A boundary surface is created. Gas bubbles that are below the boundary surface will not enter the T-piece, but will be carried further along the pipeline.

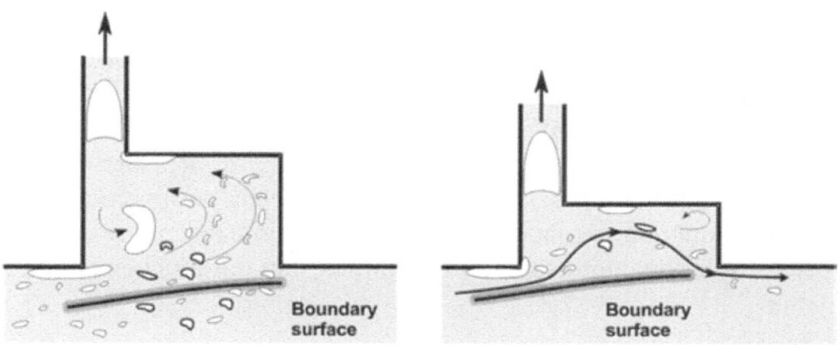

Figure A.55 Flow pattern in a T-piece with H > 0.3 D (left) and flow pattern in a T-piece with H < 0.3 D (right) (image from Wickenhauser (2008)).

Gas bubbles that are above the boundary surface will enter the lateral branch due to the circulation in the T-piece. Here the bubbles will circulate until they reach the air valve. To achieve the maximum effect of this circulation, the air valve must be mounted as far as possible on the upstream side of the lateral branch. If this is not done, it will have a significant effect on the venting capacity.

Figure A.55 (right) shows the flow pattern in a T-piece with H < 0.3 D. Here too, a circulation of gas bubbles develops, but much weaker than in the previous situation. A significant part of the gas bubbles enters and is transported out of the lateral branch and carried further along the pressure pipeline.

A.8.5 Release of gas by negative pressure

Negative pressure in the pipeline can cause the release of dissolved gas from the water since the saturation concentration c_s (in m^3 gas /m^3 water) depends on the absolute pressure in the pipeline according to Henry's Law. If we assume that water becomes saturated with a gas component under atmospheric pressure, the degassing flow rate Q_g becomes at the maximum degassing of the water at local pressure $P < P_{atm}$:

$$Q_g = \left(\frac{P_{atm}}{P} - 1 \right) \cdot c_s \cdot Q_w \tag{A.70}$$

The equation above shows the volumetric gas flow rate at the prevailing pressure P in the pipeline.

Upon its initial approach and at atmospheric pressure, the non-treated wastewater will only be saturated with nitrogen. If we assume that the wastewater is saturated with nitrogen (see Table A.8) and that all the nitrogen is released into the pipeline, then we can calculate the volume of the released gas.

Table A.8 Saturation concentrations c_s for nitrogen (N_2) at atmospheric pressure.

Temperature (°C)	Saturation Concentration (10^{-2} kg/m^3)	Saturation Concentration (10^{-2} m^3/m^3)
0	2.94	2.35
10	2.33	1.86
20	1.93	1.54

The oxygen concentration is by definition highly under-saturated. Relatively slow, biochemical processes produce all other gasses, such as carbon dioxide (CO_2), methane (CH_4) and hydrogen sulphide (H_2S). CO_2 and H_2S are very soluble in water and therefore will hardly be present in gas bubbles. Household wastewater contains insufficient sulphates (~100 mg/l) to saturate wastewater with H_2S at atmospheric pressure. Methane is even far less soluble and can , therefore, be released in gas bubbles, but practical experience shows that biochemical production is practically never the main cause of gas bubbles that reduce capacity. Only in situations with specific discharges, can the biochemical production cause gas accumulation.

Appendix B

Description of gas bubble detection methods

This appendix is based on detection methods described in the dissertations of Lubbers (2007) and Pothof (2011). If a pipeline system contains one or more gas bubbles, the stationary and dynamic character of the system change. By analysing certain signals over time, we can determine whether there is a gas bubble in the system. There are three methods available for this:

1. Changed flow rate pattern during a pump start
2. Pressure changes when draining a certain volume
3. Changes in the duration of pressure surges

B.1 FLOW RATE PATTERN DURING A PUMP START

The flow rate monitoring of a starting pump can be used as a simple diagnostic test to detect gas bubbles in the pipeline. This simple diagnostic test is mainly useful because the flow rate monitoring for each pumping station is normally available. However, this provides no information about the size and location of the gas bubble.

This method is effective because the gas bubble in the pipeline splits into two parts. The result is that the pump does not have to prime the liquid column of the entire pipeline length, but only the part up to the gas bubble. That means that the mass inertia of the liquid column is smaller, with the result that it undergoes a greater acceleration (Figure B.1).

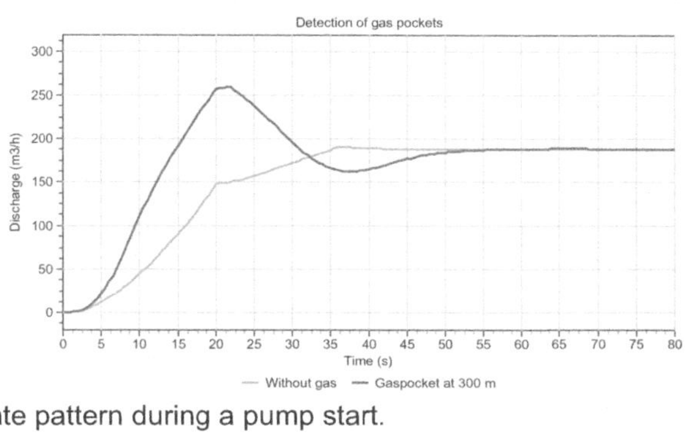

Figure B.1 Flow rate pattern during a pump start.

By the way, this method does not provide a clear result if the gas bubble in the system is very small or if the gas bubble is located at the end of the pipeline.

This method is also partly dependent on the accuracy and responsiveness (speed) of the flow rate meter. In practice, this can be a problem because most of the time electromagnetic flow rate meters are used that are quite accurate but have a low response.

B.2 PRESSURE CHANGE AFTER DRAINING

A second method of detecting gas bubbles is by registering the change in pressure when draining a certain volume. This method is based on the compressibility of liquids, the presence of gas bubbles and the rigidity of the pipeline material.

The compressibility of the liquid and the rigidity of the pipeline are defined by the water hammer storage λ (Annex A.8):

$$\lambda_0 = \frac{\Delta V}{\Delta H} = \frac{g \cdot V}{c^2} \tag{B.1}$$

This measuring method is elaborated below.

The actual measurement is done in an isolated system. Figure B.2 illustrates a simple example. By shutting two valves, the pipeline is isolated from the environment. Using a tap, a certain volume can be drained from the system, which decreases the pressure in the pipeline.

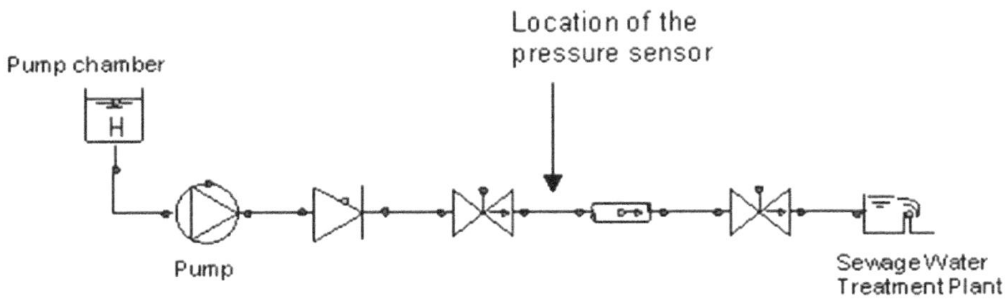

Figure B.2 The pressure pipeline can be isolated from the system by closing two valves.

If we know the drained volume and pressure decrease, then we can determine the gas volume in the pipeline. After rearranging Equation (B.1), the gas volume V_i reads:

$$V_i = \frac{p_i}{\rho g} \frac{\Delta V}{\Delta H} - \frac{gV}{c_0^2} \tag{B.2}$$

The initial pressure p_i in equation (B.2) is an absolute pressure in Pa (atmospheric). This pressure value can be obtained by adding the atmospheric pressure to the measured pressure at t=0s. Figure B.2 shows the pressure signals downstream of the valve. As a result of an expanding gas bubble, the pressure drop with a gas bubble over time is lower than the pressure drop without the gas bubble over time.

By using an average value over time (after the volume has been drained), one can determine the exact pressure drop. Figure B.3 shows that for t > 240 s the pressure has been stabilised (with the exception of water hammer surges).

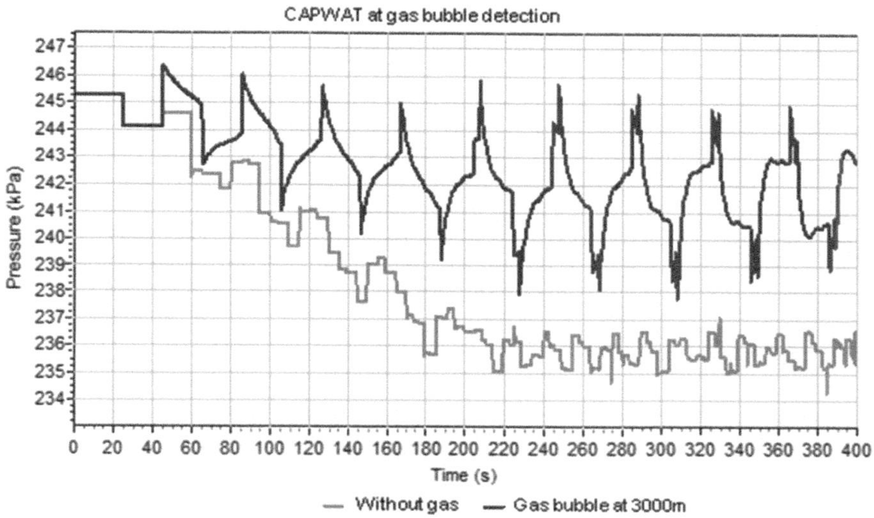

Figure B.3 Pressure signals in the system with and without a gas bubble.

B.3 DYNAMIC MEASUREMENTS

A third reliable method to detect gas bubbles in the system is to use reflections of pressure surges. By introducing a pressure surge into the system (for example, by closing a valve) and measuring the pressure, we can obtain the dynamic characteristic of the system. This dynamic characteristic is subsequently compared to the baseline-measurement ('benchmark' that was carried out during the delivery of the system) or with a simulation (using water hammer software).

The pressure wave speed of the system determines the quantity of water hammer storage. If there are gas bubbles in the system, then there is more water hammer storage than normal, which results in a lower propagation speed of the pressure wave (see Annex A.8):

$$\frac{1}{c_{gas}^2} = \frac{1}{c_0^2} + \frac{\rho \cdot V_i}{k \cdot p_i \cdot V} \tag{B.3}$$

In which:

c_{gas} = Wave speed with gas bubbles in the system	[m/s]
c_0 = Wave speed without gas bubbles	[m/s]
ρ = Density of the liquid	[kg/m^3]
V_i = Initial volume of the gas bubble	[m^3]
k = Polytropic constant	[-]
p_i = Initial absolute gas pressure	[Pa · a]
V = Total pipeline volume	[m^3]

Each system has its own basic frequency. This basic frequency is determined by the surge speed c and the characteristics of the upstream and downstream boundary conditions of the system:

For differing boundary conditions:

$$f = \frac{c}{4L} \tag{B.4}$$

For the same boundary conditions:

$$f = \frac{c}{2L} \tag{B.5}$$

in which:

f = Frequency	[Hz]
c_0 = Wave speed without gas bubbles	[m/s]
L = Length of the pipeline	[m]

In wa.stewater systems this measurement is usually carried out by closing the valve just downstream of the pumping station. These systems usually end in another receiving basin, which means that the boundary conditions differ from one another. Therefore, in this case we have to use Equation (B.4).

Therefore, there are also higher frequencies in the system for higher wave speeds. These frequencies can be made visible by means of a *Fourier-transformation*. Figure B.4 and Figure B.5 show the time series and Fourier series of a numeric model. The reference (black line) is made without gas in the system.

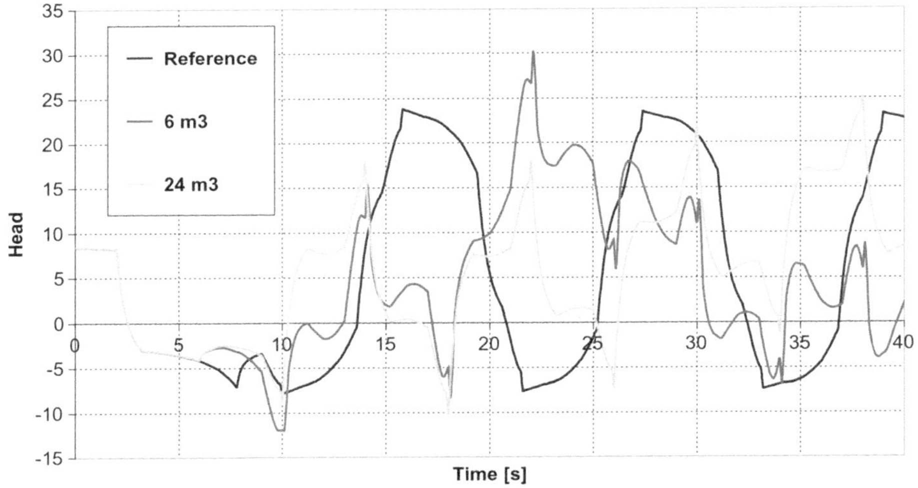

Figure B.4 Time series of several measurements with and without gas in the system (numerical example).

Figure B.5 shows the Fourier-transformed signals. The *'reference'* line clearly shows the basic frequency of 0.08 Hz (which corresponds with the duration of 12.5 s). The various lines present various gas volumes. The shift in frequency is clearly visible.

The gas volume can be derived from the shift of the basic frequency of the system. Just like the basic frequency, this depends on the boundary conditions of the system:

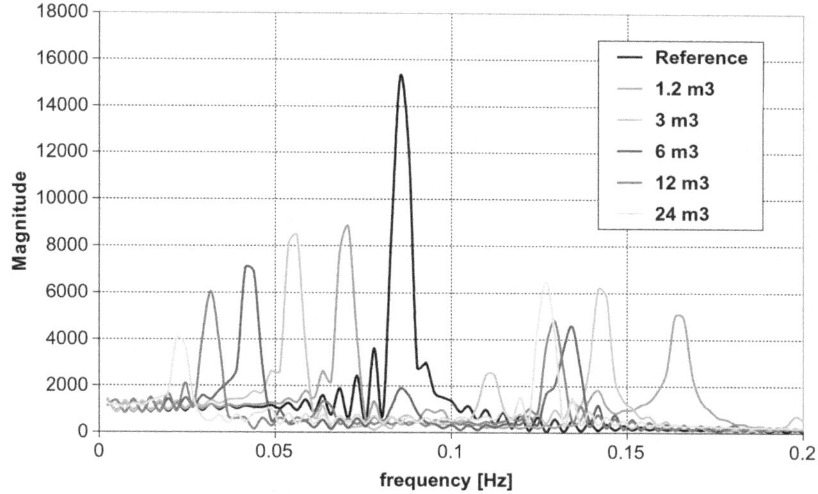

Figure B.5 Fourier series of the measurements for various gas volumes.

Differing boundary conditions:

$$V_i = \frac{k \cdot p_i \cdot V}{16L^2 \cdot \rho} \left(\frac{1}{f_1^2} - \frac{1}{f_0^2} \right) \tag{B.6}$$

Similar boundary conditions:

$$V_i = \frac{k \cdot p_i \cdot V}{4L^2 \cdot \rho} \left(\frac{1}{f_1^2} - \frac{1}{f_0^2} \right) \tag{B.7}$$

In which:

f_0 = Basic frequency without gas bubble [Hz]
f_1 = Basic frequency with gas bubble [Hz]

Figure B.5 shows a second frequency peak f_2 (well visible for V_{gas} = 24 m^3 at f_2 = 0.13 Hz). Using this second frequency, the location of the gas bubble can be determined.

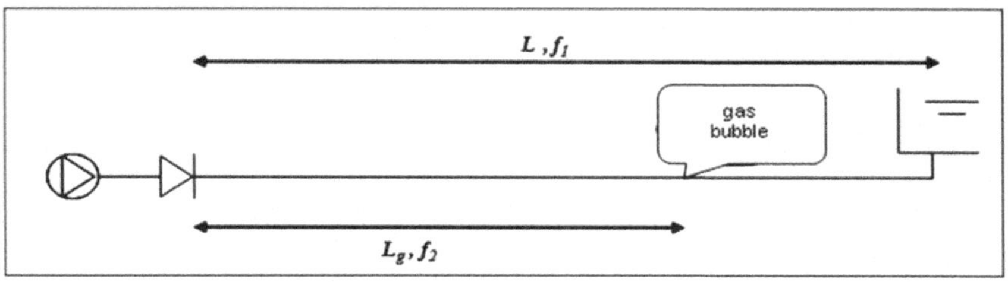

Figure B.6 Determining the location of the gas bubble.

We can determine the location of the first gas bubble by using Equations (B.4) or (B.5). The detection method is summarised in the following steps:

1. Construct a hydraulic simulation model of the wastewater transport system. The model must contain sufficient information to be able to carry out dynamic calculations (water hammer calculations). If possible, validate this model against the measured data of the system without air bubbles, so that it can be properly established that the reference signal is properly simulated. This validation can be carried out during commissioning of the system.
2. Any water hammer surge protection must be switched off in the model. While measuring in the field, the surge protections must also be switched off.
3. Determine which scenario without surge protection can be executed without damaging the system. This means that the maximum flow rate must be determined with the corresponding valve position or pump speed. Preferably, the dynamics should be initiated by closing (isolating) a valve, immediately followed by a pump stop, because this will create a well reproduced stable pressure oscillation in the pressure pipeline. Cavitation and local degassing must not occur during this scenario, because cavities and degassing bubbles cause reflections that are similar to those of the air bubbles. For this purpose, the minimum pressure in the system must remain greater than −5 m negative pressure, because approximately at this pressure level degassing starts to occur.[1] During this scenario, the dynamic pressure at the beginning of the pressure pipeline is the reference signal. When a new pipeline is put into operation, the reference pressure signal can also be measured after the system is carefully filled and all air is released; this measurement will then serve as the dynamic characteristic of the system.
4. Install a pressure gauge at the beginning of the pressure pipeline in an accessible place. An obvious location is in the pumping station, downstream from the check valves and the valve that will be closed (isolated).
5. Start the dynamic scenario in the actual system and measure the dynamic pressure. The sample time step must be at least 10 times smaller than the period of the pressure signal, potentially with an air bubble in the first crossing pipe. Suppose the first crossing pipe is at Z metres from the pumping station and the propagation speed is c [m/s]. Then the period is $4 \cdot Z/c$. The sample time T_s must be smaller than $T_s < 0.4 \cdot Z/c$ or the measuring frequency $f_m > 2.5 \cdot c/Z$.
6. The measurement signal might contain a certain trend (see Figure B.7 for example). Remove this trend. This will make the low frequency peaks in the Fourier transformation far better visible. The trend can usually be determined with sufficient accuracy by using a polynomial function (2nd or 3rd degree) or exponential function.
7. Extend the time series up to the next power of 2. The time series must be supplemented with zeroes. The trend had already been removed. There are several reasons for supplementing the time series with trivial values:

    1. The estimate of the occurring frequencies becomes more accurate because the frequency resolution equals $1/T_{hor}$ in which T_{hor} the length of the time series in seconds.
    2. With a power of 2 as the number of observations in the time series allows us to use the most efficient implementation of the Fast-Fourier Transformation (FFT).
    3. Certain tools, such as Excel, require the length of the series to be in a power of 2.

[1]The transient scenario is a balance between these two criteria. On the one hand, the pressure surge should be as large as possible to obtain a maximum amount of information and, on the other hand, the pressure surge may not be so large that cavitation, degassing or damage of the pipeline occurs.

8. Calculate the FFT of the measured pressure signal and the FFT of the reference pressure signal (calculated or measured). The FFT algorithm is available in each toolbox for time series analyses such as, for example, the *Data analysis* Add-in in Microsoft Excel. The FFT is a series of complex numbers with resolution Δf. What is known as the absolute value of the complex FFT values is used further. If $(x_j, i \cdot y_j)$ is a complex number in the FFT, the $(x_j^2 + y_j^2)^{1/2}$ is the absolute value of $(x_j, i \cdot y_j)$. This function too is standard available in Excel, Matlab or other time series analysis packages.
9. Set the absolute values of both FFT series in a graph and determine the greatest amplitude in both series. Furthermore, determine the next greatest amplitude in the measured series. The frequency with the greatest amplitude in the reference FFT is f_0, the frequencies in the greatest and the next greatest amplitude are f_1 and f_2.
10. Fill in the obtained frequencies f_2 in equation (B.4) in order to obtain the estimate of the distance to the first air accumulation.
11. From calculation model, determine the expected final pressure p_f at this location based on the reference simulation.
12. Subsequently, use the Equation (B.6) to obtain an estimate of the total air volume in the pressure pipeline.

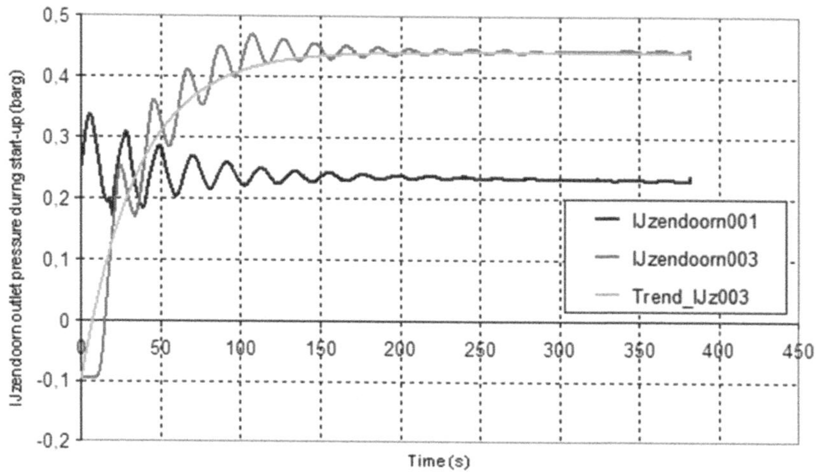

Figure B.7 Examples of dynamic pressure measurements with and without trend.

Two parameters in Equation (B.6) are not yet discussed: 1) the characteristic volume, V, and 2) the characteristic pipeline length, L. In a simple pipeline these parameters are simply the total volume and the total pipeline length. In an extended system, a reasonable estimate is obtained if the length and the volume of the main are used.

IWA Publishing's authorised EU representative for General Product Safety Regulations is Diane D'Arras, 15 rue Duret, 75116 Paris, France, e-mail: safety@iwap.co.uk.

Printed and bound by CPI Group (UK) Ltd, Croydon, CR0 4YY
12/05/2026
02108866-0009